Environmental Footprints and Eco-design of Products and Processes

Series editor

Subramanian Senthilkannan Muthu, SgT Group and API, Hong Kong, Hong Kong

This series aims to broadly cover all the aspects related to environmental assessment of products, development of environmental and ecological indicators and eco-design of various products and processes. Below are the areas fall under the aims and scope of this series, but not limited to: Environmental Life Cycle Assessment; Social Life Cycle Assessment; Organizational and Product Carbon Footprints; Ecological, Energy and Water Footprints; Life cycle costing; Environmental and sustainable indicators; Environmental impact assessment methods and tools; Eco-design (sustainable design) aspects and tools; Biodegradation studies; Recycling; Solid waste management; Environmental and social audits; Green Purchasing and tools; Product environmental footprints; Environmental management standards and regulations; Eco-labels; Green Claims and green washing; Assessment of sustainability aspects.

More information about this series at http://www.springer.com/series/13340

Subramanian Senthilkannan Muthu
Editor

Energy Footprints of the Bio-refinery, Hotel, and Building Sectors

Springer

Editor
Subramanian Senthilkannan Muthu
SgT Group and API
Hong Kong, Hong Kong

ISSN 2345-7651 ISSN 2345-766X (electronic)
Environmental Footprints and Eco-design of Products and Processes
ISBN 978-981-13-2465-9 ISBN 978-981-13-2466-6 (eBook)
https://doi.org/10.1007/978-981-13-2466-6

Library of Congress Control Number: 2018953586

This Springer imprint is published by the registered company Springer Nature Singapore Pte Ltd.
The registered company address is: 152 Beach Road, #21-01/04 Gateway East, Singapore 189721, Singapore

This book is dedicated to:

The lotus feet of my beloved Lord
Pazhaniandavar
My beloved late Father
My beloved Mother
My beloved Wife Karpagam and Daughters-
Anu and Karthika
My beloved Brother
Everyone working in the bio-refinery, hotel
and building sectors to make it
ENVIRONMENTALLY SUSTAINABLE

Contents

Energy Footprint of Biorefinery Schemes

Sara Bello, Gumersindo Feijoo and Maria Teresa Moreira

Abstract Biorefineries are evolving systems that have great potential to replace traditional oil-based alternatives. The concept of biorefinery addresses a comprehensive approach to the manufacture of bio-products and bioenergy. The intrinsic objective of a biorefinery is not to exclusively produce a single value-added bio-product such as cellulose, bioethanol, furfural, hydroxymethyl furfural, etc. The overall aim is to achieve a multi-product system with the flexibility to handle and transform different feedstocks. Different configurations evaluate the treatment of food and feed crops (first generation biorefinery), lignocellulosic biomass (second generation biorefinery) and algae (third generation biorefinery). The aim of this study is to assess the state of the art in terms of Life Cycle Assessments of biorefineries and to discuss the impact of energy consumption on global environmental outcomes. Although there is a widespread belief that biorefineries are systems with lower environmental impacts than oil-based refineries, they are energy-intensive systems with high electricity, steam and heat requirements. Therefore, a common hotspot for biorefining processes is energy consumption. The present study highlights the discussion of concepts such as the energy consumption profile of biorefineries with the aim of determining the sections of the biorefinery that could potentially contribute with higher burdens to the energy footprint of the plant. On the other hand, the evaluation of different biorefinery schemes with different functions depending on the products, raises the need to introduce concepts such as eco-efficiency to allow the comparability of the energy footprint of different scenarios. In the current framework, in which most biorefineries are pilot plants that aim to demonstrate the technical feasibility of the process under development, it is also relevant to consider aspects of energy integration and optimization. Under this

S. Bello · G. Feijoo · M. T. Moreira (✉)
Department of Chemical Engineering, Universidade de Santiago de Compostela,
Santiago de Compostela, Spain
e-mail: maite.moreira@usc.es

S. Bello
e-mail: sara.bello.ould-amer@usc.es

G. Feijoo
e-mail: gumersindo.feijoo@usc.es

S. S. Muthu (ed.), *Energy Footprints of the Bio-refinery, Hotel, and Building Sectors*,
Environmental Footprints and Eco-design of Products and Processes,
https://doi.org/10.1007/978-981-13-2466-6_1

perspective, future research has room for improvement in terms of energy use. The underlying concept is to analyze the current framework for biorefinery industries and establish benchmarks to address future research and implementation of eco-friendly alternatives. The present study suggests that industrial implementation of biorefineries in real scale systems should come with far more optimization for the achievement of sustainability. Specifically, the production of energy to fulfill the biorefinery's demand can be highlighted as one of the processes that represent clear environmental burdens. Also, pre-treatment of lignocellulosic feedstock, due to the recalcitrant nature of the biomass, can be pinpointed as an area of improvement towards the minimization of the biorefinery's energy footprint.

Keywords Biorefinery · Eco-efficiency · Energy footprint · Life cycle assessment Lignocellulosic biomass · Second generation biorefinery · Sustainability

AC	Acidification
AD	Abiotic depletion
AETP	Aquatic ecotoxicity potential
ALO	Agricultural land occupation
AP	Acidification potential
CAPs	Selected criteria air pollutants
CC	Climate change
CED	Cumulative energy demand
CED-F	Cumulative energy demand, fossil
CED-T	Cumulative energy demand, total
CHP	Combined heat and power
EC	Ecotoxicity
EIP	Exergy improvement potential
EP	Eutrophication potential
EROEI	Energy return on energy invested
EROI	Energy return on investment
EU	Eutrophication
FD	Fossil depletion
FDCA	Furandicarboxylic acid
FE	Freshwater eutrophication
FEC	Fossil energy consumption
FER	Fossil energy ratio
FET	Freshwater ecotoxicity
FEU	Fossil energy use
FU	Functional unit
GHG	Greenhouse gas
GVA	Gross value added
GWP	Global warming potential
HHC	Human health cancer
HHNC	Human health non-cancer
HMF	Hydroxymethyl furfural

HT	Human toxicity
HT-C	Human toxicity, cancer
HT-NC	Human toxicity, non-cancer
HTP	Human toxicity potential
ILUC	Indirect land use change
IR	Ionizing radiation
LCA	Life cycle assessment
LCB	Lignocellulosic biorefinery
LHV	Low heating value
MD	Minerals depletion
ME	Marine eutrophication
MEC	Marine ecotoxicity
ME-Plim	Phosphorous-limited marine eutrophication
MET	Marine ecotoxicity
MOO	Multi objective optimization
NEG	Net energy gain
NER	Net energy ratio
NEV	Net energy value
NLT	Natural land transformation
NRE	Non-renewable energy
NREU	Non-renewable energy used
OD	Ozone depletion
ODP	Ozone layer depletion potential
PA	Polyamide
PE	Polyethylene
PEG	Polyethylene glycol
PET	Polyethylene terephthalate
PHA	Polyhydroxyalkanoate
PLA	Polylactic acid
PM	Particulate matter
PMF	Particulate matter formation
PO	Photochemical oxidation
POCP	Photochemical oxidant potential
POF	Photochemical oxidant formation
POP	Photochemical oxidation potential
PS	Polystyrene
PVA	Polyvinyl alcohol
PVC	Polyvinyl chloride
REU	Renewable energy used
SED	Specific energy demand
SMG	Smog formation
SS	Subsystem
TA	Terrestrial acidification
TCF	Total chlorine-free

TET	Terrestrial ecotoxicity
TETP	Terrestrial ecotoxicity potential
TOPO	Trioctylphosphine oxide
TRL	Technology readiness level
ULO	Urban land occupation
WC	Water consumption
WD	Water depletion
WS	Water scarcity

1 Introduction. The Biorefinery Concept

The foreseeable depletion of fossil fuels demands a change in the present productive and economic structure. The development of an alternative scheme has been proposed with a view to reducing finite availability fossil resources in favor of renewable biological resources. The European Commission has set ambitious targets for reducing greenhouse gas emissions by 20% and, in parallel, increasing the use of renewable energy and energy efficiency (European Commission 2018). Within this framework, the concept of biorefinery emerges as an alternative to oil-based refineries, which requires the development of new processes through research, pilot plants and exploitation on an industrial scale (Elvnert 2009). An increasing proportion of chemicals, plastics, fuels and electricity are expected to come from biomass in the forthcoming decades. Because of its broad scope and the different drivers behind it, the sustainability of bioeconomy is expected to address important challenges in relation to social, economic and environmental aspects.

Moving from philosophy to practice, biorefineries integrate processes that convert a single biomass source into a range of biochemical materials (chemicals, materials), biofuels and bioenergy (power, heat). The core idea of a biorefinery is analogous to that of oil refineries, being both multi-product systems. Biorefineries however should engage in considering sustainability criteria, in order to compensate for low efficiencies in biomass conversion processes (King 2010).

The history of the existing corn wet-milling industry can be seen as an example of how the biorefinery of the future will evolve. Initially, the corn wet milling industry produced starch as the main product. As technology developed and the need for higher value products fostered the growth of the industry, the product portfolio expanded from starch derivatives such as glucose and maltose syrups to high fructose corn syrup. Subsequently, fermentation products derived from starch and glucose such as citric acid, gluconic acid, lactic acid, lysine, threonine and ethanol were included in the production scheme. Many other by-products such as corn gluten, corn oil, corn fiber and animal feed are currently being produced. Refineries based on lignocellulosic biomass are undergoing a similar evolution in which the product portfolio is expanding from basic wood fractions (lignin, hemicellulose and cellulose) to the production of higher value added bioproducts (mainly ethanol, but also chemicals such as furfural, hydroxymethyl furfural or furandicarboxylic acid).

In this context, there are increasing examples of biotechnology-based chemicals and materials: ethylene and isobutanol, polymers such as polylactic acid (PLA), polyethylene (PE), polyhydroxyalkanoate (PHA), enzymes, flax and hemp-reinforced composites, all of which are produced from biological feedstocks. The field of biorefinery opens up opportunities to study the environmental sustainability of processes and the relevance of environmental impacts with respect to petrochemical alternatives. Without losing the perspective of technological viability, it is necessary to address the environmental assessment of these developing processes. With this objective in mind, the consideration of the energy consumption profiles of biorefineries will make it possible to determine whether biorefineries will play a significant role in achieving the Horizon 2020 climate and energy goals. Figure 1 presents a general overview of the biorefinery approach, considering multiple products, different feedstocks and a wide range of technologies (Kamm and Kamm 2004).

A simplified comparative analysis of the basic principles of both oil and biomass refineries makes it possible to identify as a differentiating element the previous stages for the conditioning of raw materials. The petrochemical industry works on the principle of generating simple and well-defined pure products from hydrocarbons in refineries. This principle can be transferred and extrapolated to biorefineries (Fig. 2).

The aim of this chapter is to establish a basic roadmap for biorefineries under the perspective of the energy footprint. First, a review of bibliographic studies was carried out to address the state of the art in the environmental assessment of biorefineries and to analyze how the energy aspect has been described. On the other hand, an overview of biorefinery facilities in Europe has been approached with the aim of analyzing, at first hand, the state of the art on built or planned facilities. Secondly, an industrial case study has been assessed according to the life-cycle assessment approach focusing on the identification of critical process hot spots originated in the energy needs of the installation. Some concepts related to the energy footprint such as net energy gain, eco-efficiency and energy integration have been revised to provide a comprehensive view of the energy sustainability of biorefineries.

1.1 Biorefinery Configurations Attending to Feedstock

The value chain of a biorefinery is built around two relevant entities: the type of feedstock used and the separation process of the different products. Within the biorefinery, different types of biomass can be used for industrial purposes: energy crops and forestry biomass, agricultural food and feed, crop residues, aquatic plants, animal wastes and other waste materials including those from food and feed

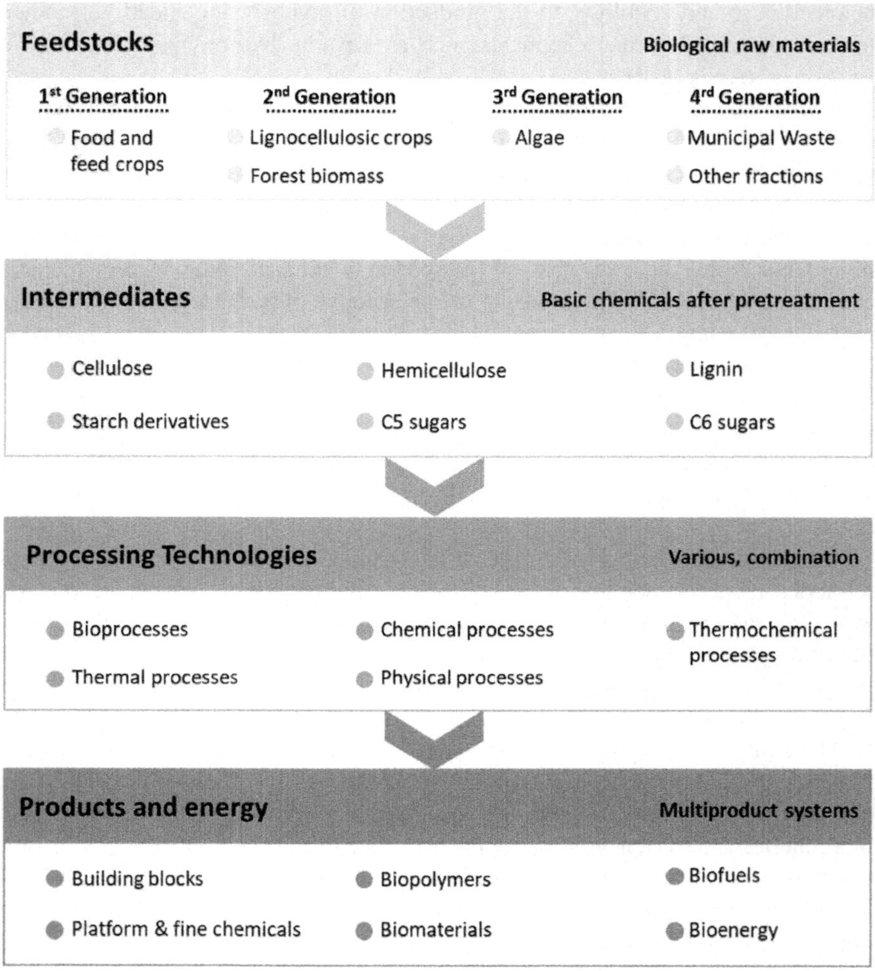

Fig. 1 Principles of a biorefinery Adapted from Kamm and Kamm (2004)

processing (Eaglesham et al. 2000). Taking the supply chain of polylactic (PLA) as an example, sugar-based biomass (e.g. sugar cane, sugar beet, etc.) is used as a substrate to obtain lactic acid or lactides. These lactides eventually form the basis of PLA, which can be sold as such and/or used to produce other consumer end products.

Some authors suggest the existence of four different biorefinery configurations that have been defined according to the type of feedstock they intend to exploit. Obviously, the biomass to be processed affects the viability of the technologies to be used in each case. Generally speaking, the exploitation and processing of bio-based feedstocks will be closely linked to the technology needs and the energy consumption of processes. In terms of potential profitability, it is important to

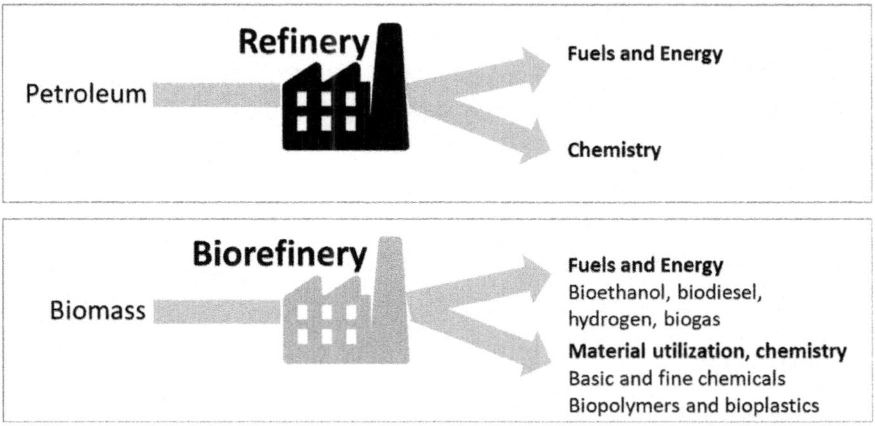

Fig. 2 Basic principles of traditional refinery and biorefinery Adapted from Kamm et al. (2008)

assess the strengths, weaknesses, challenges and opportunities that can be effectively applied to improve the prospects for a sustainable biotechnology-based economy.

Biomass production systems, supply chains and end-uses differ widely, according to different feedstocks, and so do their environmental and socioeconomic impacts (e.g. carbon stocks, water, soil, air, biodiversity, land use change and food security). The direction (positive or negative) and magnitude of these impacts mainly depend on the type of feedstock, biophysical and socio-economic conditions of the production site, production technologies, supply chain and end-use.

1.1.1 First Generation Biorefineries

The use of agricultural resources in industry was first proposed in the 18th century with the development of technologies for corn refining. This achievement marked the first step towards the evolution of the biorefinery approach. Until the conquest of the lead position by crude oil as the primary fuel in the industrialization process of the 19th and 20th centuries, extensive exploitation of biomass was mainly linked to the use of agricultural resources (Kamm et al. 2016).

Today, first-generation biorefineries are facilities that exploit edible crops such as grains, sugar, starch or oilseeds. Some of the most common food crops processed in biorefineries are maize, wheat, triticale, sorghum, rice, sugar cane, sugar beet, cassava, soybean, oil palm and rapeseed (Cassman and Liska 2007). In Europe and North America, most bioethanol is produced from maize and wheat (Vohra et al. 2014). However, it is recognized that the production of first-generation sugars implies the need for large quantities of feedstocks available at an uncompetitive price; conventional crops could not meet the potential global demands for biofuels to counteract declining fossil fuel reserves, mainly because of potential competition with food and feed markets, which also generates widespread social controversy (Sarkar et al. 2012).

Edible crops provide a high sugar content, which in turn leads to increased production yields of sugar-derived products (e.g. bioethanol). The challenge for first-generation biorefineries is to be able to exploit crops without causing potential damage to food security, arable land or land-use change (Gnansounou and Pandey 2017).

1.1.2 Second Generation Biorefineries

Agro-industrial residues, non-edible crops and forestry products present opportunities to avoid the use of food-based feedstock in biorefineries. Within the scope of second-generation biorefineries, different raw materials such as grass, straw, hemp, forest biomass or harvest residues from crops can be included (Stuart and El-Halwagi 2014).

The reuse of crops that produce woody by-products or crops not intended for food production avoids a speculative increase of food prices (Hatti-Kaul 2010). Current research trends focus on the lines of lignocelluloses and feedstocks that provide lignin, hemicellulose and cellulose. Barriers related to fractionation of lignocellulosic biomass, energy needed for product separation, biological and chemical inhibition and better integration of the entire process chain should be considered (E4tech et al. 2015). The opportunities arising from the use of unprofitable fractions of lignocellulosic biomass make it possible to increase the intrinsic value of the raw material by producing several high added value chemicals.

Second generation biorefineries go beyond the use of food as fuel. However, one of the potential challenges faced by this category of biorefineries is the potential diversion of arable land use from food production to energy production. Such is the case of energy crops, an option that avoids the use of food as a raw material for bioenergy production, but requires land-use change (Harris et al. 2015).

To avoid this concern, a conceivable option would be to transform biomass fractions that have a minimal impact on the use of water, fertilizers, herbicides, machinery, as well as land-use change. Lignocellulosic by-products or waste fractions from crop cultivations that would have no other application are some potential examples (Tomei and Helliwell 2016).

Pretreatment of Lignocellulosic Biomass as an Essential Requirement

Within the European framework, second-generation biorefinery, also known as lignocellulosic biorefinery (LCB), uses wood (including forest residues and black liquor) and straw as main feedstocks (Biorefinery Euroview 2008). The transformation process in an LCB consists of four main steps: pre-treatment, hydrolysis, fermentation and product purification. In its natural state, lignocellulosic material is difficult to be treated by direct hydrolysis of cellulose into glucose. Therefore, the fractionation of lignocellulosic biomass is one of the most complex operations among biorefinery processes, mainly due to the structure of solid and interconnected cell walls of biomass. The complex polymer structure of cellulose and the integrated base of hemicellulose and lignin tend to obstruct and prevent its conversion into monomeric sugars (Kamm and Gruber 2006).

Based on these factors, it is necessary to develop effective pre-treatment stages to reduce the size of material particles and alter their cellular structure to make it more accessible to chemical or enzymatic hydrolysis processes (Himmel et al. 2007). These processes can be based on mechanical, physical, chemical and/or biological treatments.

The selection of the pre-treatment method plays a critical role in the transformation of lignocellulosic biomass in a viable and cost-effective way (Kautto et al. 2013). Several pre-treatment methods have been studied and in general, this step has been considered one of the most costly processes in the conversion of lignocellulosic material (Harmsen et al. 2010). All pretreatment techniques can be classified into four different groups, as depicted in Table 1. The main objectives of pretreatment technologies are to improve the yields of hexoses and pentoses in downstream processing by ensuring lignin recovery, decrease costs in size reduction of biomass, minimize energy and chemicals requirements, be flexible enough to process different lignocellulosic feedstock and reduce waste production (Alvira et al. 2010).

Table 1 Lignocellulosic biomass pretreatments Adapted from Prasad et al. (2016)

Pretreatment category	Methodology
Physical	Wet milling
	Dry milling
	Grinding
	Microwave
Chemical	Alkaline hydrolysis
	Acid pretreatment
	Organosolv process
	Ozonolysis process
	Wet oxidation
Biological	Fungal degradation
Physicochemical	Steam explosion
	Ionic liquids
	Catalyzed steam explosion
	Ammonia fiber explosion
	Liquid hot water

Going Deep into Lignocellulosic Biorefineries: Organosolv Process

Among the pretreatment techniques for wood fractionation, organosolv pretreatment has been found to have the advantage of using solvents that can be easily recovered while obtaining high quality lignin (Alvira et al. 2010). During the process, an organic solvent mixture with inorganic acid catalysts (HCl or H_2SO_4) is used to break down the internal structure of lignin and hemicellulose. The most common organic solvents used are methanol, ethanol, acetone, ethylene glycol, triethylene glycol and tetrahydrofurfuryl alcohol (Chum et al. 1988). Organic acids can also be used as catalysts during the process; however, at high temperatures (above 200 °C), the addition of catalyst is unnecessary for delignification (Aziz and Sarkanen 1989) but leads to a high yield of xylose. Once the reaction is complete, it is necessary to recover the solvent for reuse, as it may inhibit the subsequent stages of enzymatic hydrolysis and fermentation.

Its use as a fuel for heat and electricity production are common applications of the large amounts of lignin generated in pulping processes (Kleinert and Barth 2008). Recent studies have shown that due to its high quality, organosolv lignin offers different applications as a substitute for phenolic resins or polyurethane compounds (Pandey and Kim 2011). Besides lignin, many other co-products can be recovered from the main stream of hemicellulose, including sugars, acetic acid and furfural. Cellulose and hemicellulose can be hydrolyzed enzymatically to C6 and C5 sugars. These sugar flows can be further fractionated, offering opportunities for the production of biofuels and bio-based chemicals (E4tech et al. 2015).

Although the use of organosolv as a pre-treatment may benefit the production of co-products, its practice has been assumed to be more complex and costly than other methods, due to the high energy consumption in distillation and safety costs and the potential risks of fire and explosion (Zheng et al. 2009). In views of cushioning the high costs of production of organosolv pulp, an attempt should be made to recover all possible products at subsequent stages of processing.

1.1.3 Third Generation Biorefineries

Third generation biorefineries use aquatic biomass such as algae to produce, mainly, biodiesel or vegetable oil due to their high oil content (Faraco 2013). Algae and microalgae are considered a very promising feedstock as they require CO_2 for their growth, which can counteract GHG emissions. Moreover, this feedstock does not compete directly with other crops for arable land, as it is grown in photobioreactors or raceway ponds (Gavrilescu 2014).

Algae growth rates and reactor design should be optimized to maximize production; optimized production would allow efficient conversion to protein, carbohydrates and lipids. However, the bottleneck of marine biorefinery is the harvesting and subsequent extraction. The potentiality of third-generation biorefineries is increasing, due to the multiple efforts towards technological advances, as well as the possibility of not only producing biodiesel, but also other products such as ethanol, hydrogen, liquid fuels, methane and high value products (pigments, antioxidants, carotenoids, proteins). In terms of sustainability, algae biorefineries present strengths over the feasibility of reusing nutrient-rich wastewater instead of saline water (Martín and Grossmann 2013).

1.1.4 Fourth Generation Biorefineries

Some authors propose the inclusion of an additional category of biorefineries for those systems that exploit raw materials that do not belong to any other category (Demirbas 2010; Gavrilescu 2014; Haddadi et al. 2017; Stuart and El-Halwagi 2014). In the case of fourth-generation biorefineries, the main feedstocks are waste fractions, such as municipal waste. These biorefineries follow a circular economy approach, using waste that is difficult to manage and has the potential to produce biofuel.

Fourth-generation biorefineries potentially include facilities for the treatment of feedstocks that are not directly related to crop cultivation, use of arable land or production of marine feedstock. Rather, they are intended for the valorization of waste fractions such as those from vegetable oils, food industry and even sewage sludge. These new-generation biorefineries may not follow the standard structure of a biorefinery plant and may be combined with wastewater treatment plants or industries to produce valuable products from waste and therefore manage such waste on-site (Haddadi et al. 2017). An example of the fourth-generation biorefinery concept is the production of polyhydroxyalkanoate from primary and secondary sludge in wastewater treatment plants (Morgan-Sagastume et al. 2014; Mosquera-Corral et al. 2017).

1.2 Biorefinery Configurations Attending to Products

Some biorefineries have fixed processes and produce a fixed amount of ethanol and other end products, while other configurations can produce multiple end products. The flexibility of the plant to use a blend of biomass feedstocks influences the possibility to produce a variety of products by combining technologies (Kamm and Kamm 2004).

One of the objectives of a biorefinery is to obtain products in concentrations that make purification or recovery economically feasible (Mosier et al. 2005). In fact, some authors (Boisen et al. 2009) argue that a biorefining facility should not be limited to the production of a single high value added bioproduct and that bio-based raw materials should be used as efficiently as possible.

Therefore, we can find that a wide range of bio-based products can be obtained depending on the production targets of the biorefinery and technology readiness level (TRL) of the downstream processes. On the other hand, the layout of the plant may vary depending on whether the main production objective is to obtain mainly bioenergy/biofuels or high added value products. In any way, biorefineries are viewed, in most cases as complex systems with multi-production perspectives. Not all plausible products that can be obtained from the biorefinery route have equally developed TRL, the same market size or equal potential market forecasts. Listed below are some of the possible products manufactured in biorefineries (E4tech et al. 2015).

- Basic bio-based building blocks. Lignin, hemicellulose, cellulose, glucose, fructose, galactose, xylose, arabinose, ribose, lactose, sucrose, maltose.
- Platform and fine biochemicals. Methane, formic acid, ethanol, acetic acid, glycolic acid, lactic acid, propionic acid, succinic acid, xylitol, levulinic acid, furfural, hydroxymethyl furfural (HMF), citric acid, furandicarboxylic acid (FDCA), lipids, 1,4-butanediol, ethyl acetate, cyrene.
- Biopolymers. Polyamide (PA), polyethylene (PE), polyethylene terephthalate (PET), polyethylene glycol (PEG), polyvinyl alcohol (PVA), polyvinyl chloride (PVC), polystyrene (PS), polyhydroxyalkanoates (PHA).
- Biomaterials. Foams, composites, bioplastics and films (manufactured from biopolymers).
- Biofuels/bioenergy. Gasoline, jet fuel, diesel, alkanes, biogas.

1.3 Biorefinery Under the Focus of Sustainability

Recently, several studies have performed an environmental evaluation of biorefinery systems. Although most of them have confirmed that bio-based products present lower environmental burdens than fossil-based products, a new concern is the wide range of biomass feedstock alternatives and emerging technologies for conversion, from which the most environmentally friendly should be chosen for future biorefinery processes (Stuart and El-Halwagi 2014).

Life Cycle Assessment (LCA) is a methodology for assessing the potential environmental impacts and resources consumption associated with a product or production system throughout its life cycle, as well as identifying opportunities for environmental benefits (ISO14044 2006a, b). Different environmental assessment studies have been carried out on biorefinery systems. However, it is difficult to compare their results because they have considered different feedstocks, technology treatments, system boundaries or methods of environmental allocation.

González-García et al. (2011) identified and quantified the environmental impacts associated with a Swedish softwood-based biorefinery where total chlorine-free (TCF) cellulose, ethanol and lignosulfonates were produced. They have found that the production of chemicals consumed in the cooking and bleaching stages, the treatment of sludge generated in the wastewater treatment plant and the on-site energy production system were the elements that contributed most negatively to environmental burdens. Hernández et al. (2014) studied an olive stone based biorefinery and carried out an environmental assessment of two biorefinery schemes describing the integrated production of xylitol, furfural, ethanol and a cogeneration system to produce bioenergy from solid waste. The results showed that for both biorefinery schemes, there were considerable net profit margins. Regarding the environmental analysis, they concluded that the cogeneration system reduced energy consumption.

Laure et al. (2014) assessed an organosolv lignocellulose biorefinery at pilot plant scale, highlighting the benefits of a lignocellulose biorefinery and the importance of valorizing all the fractions obtained in order to create a competitive bio-production. Budzinski and Nitzsche (2016) evaluated four conceptual beech wood based biorefineries. The results indicated that the four biorefinery systems had fewer total potential environmental impacts than fossil-based reference systems. González-Garcia et al. (2016) highlighted the relevance of multi-product valorization when considering the environmental performance of biomass refining into high-added value compounds.

1.4 Energy Security

The European Commission, on the Energy 2020 Strategy (European Commission 2010), defines energy security as the uninterrupted physical availability of energy products on the market at an affordable price for private and industrial consumers, while contributing to the EU's social and climate objectives. Europe energy policies base the main objectives to be achieved by 2020 on ensuring security of energy supply, competitiveness and sustainability. The sustainability objective is based on the development of environmental quality systems that produce energy from renewable sources. The concept of energy security can therefore be closely linked to the sustainability of biorefinery systems.

In the case of biorefineries producing biofuels, sustainability is addressed, for example, by exploiting feedstocks mentioned in Sect. 1.1. Moving towards a biotechnology-based economy is an opportunity to achieve the established targets for energy security in Europe. The strategic objectives provide an alignment towards decarbonizing energy sources, giving priority to renewable energies, supplying and using energy efficiently and improving energy technologies and innovations (European Commission 2010).

2 Life Cycle Assessment of Biorefinery Schemes

Life cycle assessment is a tool that has been widely used to report environmental sustainability criteria of biorefineries for the production of bioenergy (Li et al. 2018) and bioproducts. Biorefineries are inherently characterized by their flexibility, as seen in Sect. 1, they offer a wide range of possibilities. Thus, the results derived from LCA may be divergent when assessing different types of facilities.

The functional unit, objective and scope, system boundaries and method of each individual study affect the overall results reported on LCA. This section aims to conduct a review of biorefinery LCA studies on literature to evaluate the overall profile of different biorefinery schemes and to assess the relative implications on issues such as the relevance of energy footprint. The state of the art regarding the life cycle assessment of biorefineries can be described through the sample of studies presented in Table 2. The sample includes 31 peer-reviewed papers that are considered representative of research from the last decade.

2.1 Goal and Scope Definition

The definition of the goal and scope in LCA should be clearly stated, providing the intended application and reasons for conducting the assessment, the functional unit, the system boundaries and inventory data (ISO14044 2006a, b).

Among the reviewed papers, 52% were studies on second generation biorefineries, 24% on first generation biorefineries, 18% on third generation biorefineries and finally 6% on fourth generation biorefineries. This clearly indicates that the research trends have been focused on the valorization of lignocellulosic materials. Second generation biorefineries take a relevant share, lower however than first generation biorefineries, which are far more implemented industrially on a real scale. Studies on valorization of algae and municipal solid waste are far from being adopted industrially since they are in the early stages of technological development at laboratory or pilot scale.

In terms of products of interest for each configuration, the conclusion is that biorefineries tend to be more sustainable or economically viable, either in the production of energy and biofuels or energy/biofuels together with bioproducts. Of the documents reviewed, 68% considered the production of biofuels and/or electricity along with one or more bioproducts. In contrast, 19% and 13% considered the production of biofuels/bioenergy or of bioproducts exclusively, respectively.

2.1.1 Functional Unit

The functional unit (FU) provides a reference to which input and output data are normalized, it should be mathematically measurable. The functional unit should be carefully selected to allow comparisons between the valorizing systems under study (ISO14044 2006a, b). Attention should be paid to the selection of FU since decision-making strategies may depend heavily on it. Based on the literature review, one quarter of the studies have selected a FU referred to the feedstock, while the remaining three quarters have chosen a FU related to the products.

Feedstock-based functional units include volumetric or mass values (16% of reviewed documents) and hectares of land (9%). Among product-based FU, the variability is greater. 35% of the papers have chosen a FU that represents the

Table 2 Literature review of LCA biorefinery studies

References	Feedstock	Products	Functional unit	System boundary	Data quality	Method	Environmental categories/ indicators	Energy indic.
Adom and Dunn (2017)	Corn Corn stover Glucose	Polymer grade lactic acid Ethyl lactate	kg product	C-G, C-GR	2, 3, 4	GREET	GHG	FEC
Arodudu et al. (2017)	Maize	Ethanol Biogas	ha of land	C-G	2, 3	–	–	NEG EROEI
Barlow et al. (2016)	Algae	Biofuel	1 MJ of fuel	W-P	2,3,4	–	GWP	NER
Collet et al. (2014)	Microalgae	Methyl ester Glycerin	1 MJ methyl ester	C-G	2	ReCiPe	CC, OD, HT, FET, MET, POF, PMF, IR, FET, ME, FE	CED
Escamilla-Alvarado et al. (2017)	Municipal solid waste	Hydrogen Methane Holocellulases Hydrolysates	1 ton municipal solid waste (20% total solids)	G-G	1, 2, 3	CML 2001	AD, AP, EP, GWP_{100}, OD, POP	–
Farzad et al. (2017)	Sugarcane residues	Ethanol Lactic acid Furfural Butanol Methanol Electricity	65 tons residues/h processed at the biorefinery	C-G	2, 3, 4	CML-IA baseline	AD, GWP, AC, EU, ODP, POCP, HT, FET, TEC, WC, WS	SED
Fiorentino et al. (2014)	Brassica Carinata	Ethyl levulinate Formic acid Glycerin Biodiesel Lignin Electricity	1 ha of marginal land with B. carinata	C-G	1, 2, 3	CML 2001	AD, AC, EU, GWP, HT, PO	CED

(continued)

Table 2 (continued)

References	Feedstock	Products	Functional unit	System boundary	Data quality	Method	Environmental categories/ indicators	Energy indic.
Giwa (2017)	Algae and Cattle manure	Biogas Energy	1 GJ biogas/year	C-GR	2, 3	Eco-indicator 99	Carcinogens, respiratory organics and inorganics, radiation, ecotoxicity, land use, minerals, fossil fuels, CC, OD, AC, EU	–
Gnansounou and Raman (2016)	Algae	Biodiesel Glycerol Animal feed Succinic acid	1 km	W-WH	2	ReCiPe	GHG, land use	FEU
Gontia and Janssen (2016)	Pulp mill residual streams	Sodium Poly-acrylate Ethanol	1 kg sodium poly-acrylate	C-G	1, 2, 3	CML	GWP, EP, AP, POCP	REU NREU
González-García et al. (2011)	Softwood	Lignosulfonates Ethanol	1 ton air-dried (10% moisture) dissolving cellulose from pine and spruce	C-GR	1, 2, 3	CML-2 baseline 2000	AD, GW, OD, HT, FE, ME, TE, PO, AC, EU	–
González-García et al. (2016)	Residual wood	Cellulose Lignin Monosaccharides Hemicelluloses Levulinic acid Formic acid	1 € revenue	C-G	1, 2, 3	ReCiPe	CC, OD, TA, FE, ME, HT, POF, TEC, FEC, MEC, WD, FD	–

(continued)

Table 2 (continued)

References	Feedstock	Products	Functional unit	System boundary	Data quality	Method	Environmental categories/ indicators	Energy indic.
Khoshnevisan et al. (2017)	Castor	Biodiesel Ethanol Electricity Methane Glycerol K_2SO_4	1 GJ output energy from combustion of castor biodiesel blend	W-WH	2, 3	IMPACT 2002+	Resources Human Health Ecosystem quality	NEG FER
Kim and Dale (2015)	Corn stover	Ethanol Electricity	1 MJ ethanol	C-G	2, 3	GREET	GW, AC, EP, SMG, OD, EC, PM, HHC, HHNC	FEU
Levasseur et al. (2017)	Hydrolysate in Kraft pulp mill	Ethanol Butanol Acetone	1 kg butanol	C-G	2, 3	IMPACT 2002+	Resources Human health Ecosystem quality	–
Lin et al. (2015)	Lignocellulosic Biomass	Glucose Hemicelluloses HMF Levulinic acid 2,5-hexanedione Xylene	1 ton p-xylene	C-G	2, 3, 4	ReCiPe	OD, HT, EC, IR, POF, PMF, CC, TET, TA, ALO, ULO, NLT, MET, ME, FE, FET, FD, MD, WD	–
Mandegari et al. (2017)	Bagasse Brown leaves	Glucose Xylose Electricity Lactic acid Ethanol	1 ton lactic acid	C-G	2, 3, 4	CML-IA Baseline	AD, GWP_{100}, AC, EU, OD, POCP, TET, FET, HT	SED
Modahl et al. (2015)	Timber Wood chips	Ethanol Cellulose Lignin Vanillin	1 ton product 1 m^3 ethanol	C-G	2, 3	CML 2 Baseline 2000 IPPC 2007	CC, AC, EU, POF, OD	CED

(continued)

Table 2 (continued)

References	Feedstock	Products	Functional unit	System boundary	Data quality	Method	Environmental categories/ indicators	Energy indic.
Moussa et al. (2016)	Sorghum	Succinic acid $(NH_4)_2SO_4$	1 kg succinic acid (99.5 wt%)	C-G	1, 2, 3	IPPC 2007	GWP_{100}	CED SED
Ofori-Boateng and Lee (2014)	Oil palm fronds	Ethanol Phytochemicals	1 ton anhydrous ethanol (99.15 wt %)	C-G	2, 3	CML 96	AP, AETP, EP, GWP, HTP, ODP, POCP, TETP	EIP
Olofsson et al. (2017)	Agricultural crop residues	Ethanol Electricity Lignin	1 MJ bioethanol (LHV)	C-G	2, 3	Renewable energy directive (RED)	GWP	Primary energy
Parajuli et al. (2017)	Wheat straw Alfalfa	Ethanol Lactic acid Electricity	1 MJ ethanol (99.5 wt%) and 1 kg lactic acid	C-G	2, 3	ReCiPe	GWP_{100} with and without ILUC, EP, ALO	NRE
Pereira et al. (2015)	Sugarcane	Ethanol Acetone n-Butanol Electricity Sugar	kg butanol US$ earned km run by car	C-G	2, 3, 4	CML 2 Baseline 2000	AD, AC, EP, GW, HT, PO	–
Rahimi et al. (2018)	Eruca sativa	Biodiesel Electricity Ethanol Heat Glycerol Biomethane	1 GJ output energy from biodiesel	W-WH	2, 3	IMPACT 2002+	Resources, human health, ecosystem quality	NEG FER
Raman and Gnansounou (2015)	Vetiver leaves	Lignin Ethanol Furfural	1 km passenger travel operated on ethanol	W-WH	2, 3	ReCiPe	CC, FD	–

(continued)

Table 2 (continued)

References	Feedstock	Products	Functional unit	System boundary	Data quality	Method	Environmental categories/ indicators	Energy indic.
Seghetta et al. (2016)	Seaweed	Bioethanol Fertilizer Fish feed	1 ha of sea under cultivation	C-GR	1, 3, 4	ReCiPe	CC, ME, ME-Plim, HT-C, HT-NC	CED-T CED-F
Silalertruksa et al. (2015)	Sugarcane	Bioethanol Biogas Sugar Electricity	Unit of product	C-C	1, 2, 3	ReCiPe	GHG emissions, GVA	–
Souza et al. (2015)	Sugarcane Algae	Biodiesel Ethanol Algae meal Electricity	1 MJ hydrous ethanol	C-G	1, 2, 3	Monte Carlo simulation	GHG emissions	FEU
Tao et al. (2014)	Corn Stover	Isobutanol Ethanol n-Butanol	1 km travelled by a flex-fuel car operated on biofuels	F-WH	3, 4	-	GHG, EROI, CAPs	FEU NEV
Uihlein and Schebek (2009)	Straw from agriculture	Xylite Lignin Ethanol	1 kg ethanol	C-G	2, 3	Eco-indicator 99	Human health, resources and ecosystem	–
Vaskan et al. (2018)	Palm empty fruit bunches	Ethanol Animal feed Electricity	1 ton empty fruit bunches	W-WH	3, 4	ReCiPe	CC, FD, HT, FET, FE & economic indicators	–

1 Primary data, 2 literature data, 3 databases, 4 modelling/simulation, C cradle, G gate, GR grave, W well, WH wheel, F field, P product

quantity of product, either in mass or volumetric values, with units such as kg, ton, m^3, L. Close follows 28% of the papers that have opted to use a FU referred to energy, with units such as MJ or GJ of energy produced in the form of biofuels or electricity. The distance travelled by a car fueled on biofuel is a relatively common FU, although only 9% of the reviewed studies have worked with it. Finally, only 3% of the studies have used the monetary benefit as a reference unit.

The functional unit should not only represent a number and a unit. It involves the specific circumstances of the study under which such number and unit make sense. For instance, some of the reviewed papers refer to a time frame, geographical location or composition value.

2.1.2 System Boundaries

The system boundary describes all the unitary processes included within the system evaluated through LCA. The stages and boundaries selected for the study, as well as the omissions considered, should be identified and explained. It is helpful to describe the systems using process flows diagrams that show where the unitary process begins with raw materials and ends with the management of the final products (ISO 14044 2006a, b).

A large portion of the reviewed papers (59%) have studied the process from the feedstock production to the final products at the plant gate (cradle to gate). A slightly broader scope has been adopted in 13% of papers, including the disposal phase (cradle to grave). The well-to-wheel system boundary is a common scope used for studies related to the production of biofuels for vehicles, considered in 16% of the studies. It is important to note that the selection of system boundaries may be influenced by the availability of data on issues such as end-of-life or waste management.

2.1.3 Inventory Data

Inventory data of studies provides the basis for an environmental evaluation that may be representative of specific processes or products. Data quality and completeness of inventories influence the reliability of the life cycle assessment results. This review compiled sources of inventory data. Almost all studies consider a combination of different sources to ensure completeness of inventory gaps. For instance, when primary data is available, background process data is often implemented through literature or database information.

Regarding biorefinery systems, most of the data is obtained from literature and databases (38 and 39% respectively), being Ecoinvent one of the most common databases. Primary, pilot facility or large-scale data are very rare (11%), as is the use of inventories from modelling or simulations, which account for 12% of the data retrieved in reviewed studies.

2.2 Method and Impact Categories

The method selected to perform the environmental evaluation determines the characterization factors and model that represent the aggregated impacts of inputs and outputs. The method is usually implemented through specialized software and provides a set of impact categories as for the collective description of results (ISO 14044 2006a, b). The most commonly used methods among the reviewed papers were CML in its different versions (IA Baseline 2000, 2002) and ReCiPe. In general, both these methods were applied in two-thirds of the studies.

Among the impact categories, the most frequently used was global warming potential in all its variants (climate change, greenhouse gas emissions). In fact, studies often focused exclusively on determining this environmental category through the calculation of kg CO_2 of input and output flows. Other set of indicators that were relevant for the evaluation of biorefineries were ozone depletion, acidification, eutrophication and toxicity.

Due to the relevance of energy consumption and the feasible production of bioenergy in biorefineries, the study of energy indicators on biorefineries is relevant. Some of the revised documents have considered energy-related indicators. Of 31 studies, 20 used at least one indicator representing energy consumption or production. The most common indicators were the net energy gain (NEG), cumulative energy demand (CED), specific energy demand (SED) and fossil energy use (FEU). Overall, the evaluation of the energy footprint of biorefineries has certainly not been studied to the point of accomplishing environmental optimization of energy-related procedures.

Conclusively, LCA studies on biorefinery systems should typically address some key features and methodological assumptions. Life cycle assessment is a discipline in which the main goal is to analyze environmental performance of a process. However, many times, the study of the environmental performance of a process does not show environmental excellence by its very nature. Upon analysis of the presented overview, the goal of LCA, particularly in novel processes should not only be to provide environmental results of a process but to establish a benchmark for comparison with reference systems or with different processing routes.

The optimality of a life cycle assessment study is determined by the nature and quality of the available data. The scope of the study should cover the areas in the value chain in which data is available and reliable. If data is available, it is advised that the feedstock production is included within system boundaries, since it represents one of the most characteristic distinctives of biorefinery systems, which is the exploitation of biomass with the purpose of providing biofuels and biochemicals.

According to the literature review, the variability on the selection of functional unit is high. This, again, is highly impacted by the type of system under study. When the biorefinery system clearly has a wide range of products in their portfolio, and there is not one that can be highlighted as the main product (according to production volume, economic relevance, etc.), the suggestion on the selection of functional unit is to favor feedstock-based reference units. If the target, however, is to evaluate the energy footprint of biorefinery scenarios, it may be more appropriate to define a functional unit referred to the output energy/fuel production, for instance, 1 MJ of fuel. If the aim of the study is to describe a system up to the use of a biofuel-powered vehicle, then is coherent to adopt functional units that include the vehicle's features (e.g. efficiency), for example 1 km travelled in a specific vehicle powered by the produced biofuel.

Finally, with respect to the selection of impact categories and indicators, typically, LCA studies have followed the trend of focusing on carbon emissions. However, considering the potential nature of diverse processes in the biorefinery field, it is advised that at least one indicator is included out of categories such as ozone depletion, acidification, eutrophication and toxicity. It has been observed that energy-related indicators such as NEG, CED, SEG, etc. act as fair descriptors of biorefinery systems, in terms of sustainability, percent use of fossil resources or energetic efficiency.

3 Biorefineries in Europe. State of the Art

The theoretical vision of biorefineries in literature has been revised continuously in one or another way. As it has been mentioned, many research studies have analyzed environmentally biorefinery related processes. There are, as well, plenty of research studies that evaluate and analyze the production of biochemicals, or biofuels, mostly at laboratory scale. However, the best overall vision that one could have in the field of biorefineries, is through an analysis of existing biorefining facilities. For the purpose of this study, the search of facilities has been narrowed down to European facilities producing biofuels, bioenergy or bioproducts. Through the evaluation of common characteristics in existing biorefineries, the expected result is to obtain an overview of the state of the art and future possibilities and prospects in the field of sustainable processing.

A total of 568 biorefinery facilities were reviewed throughout Europe in available databases and compilations (Bioenarea 2010; Biorefinery Euroview and Biopol 2009; E4tech et al. 2015; IEA Bioenergy 2018; Nova Institute and Consortium 2017). In this chapter, different types of production plants were considered. The scope includes processing plants to obtain bioethanol, biodiesel, bioproducts as well as power plants that use coal and biomass blends as fuel. The first objective was to analyze which were the European countries with highest density of biomass transforming facilities. Figure 3 displays a density map featuring reviewed biorefining facilities. From the evaluated group of biorefineries, Germany was the country with more biomass processing plants, with a total of 132. Finland was found to be quite relevant as well, with 102 plants. Other countries such as Italy, Denmark, France, The Netherlands, Sweden and United Kingdom were found to have an intermediate number of facilities, ranging from 23 to 55 factories. Spain, Ireland, Hungary, Czech Republic, Norway, Poland, Portugal, Slovakia or

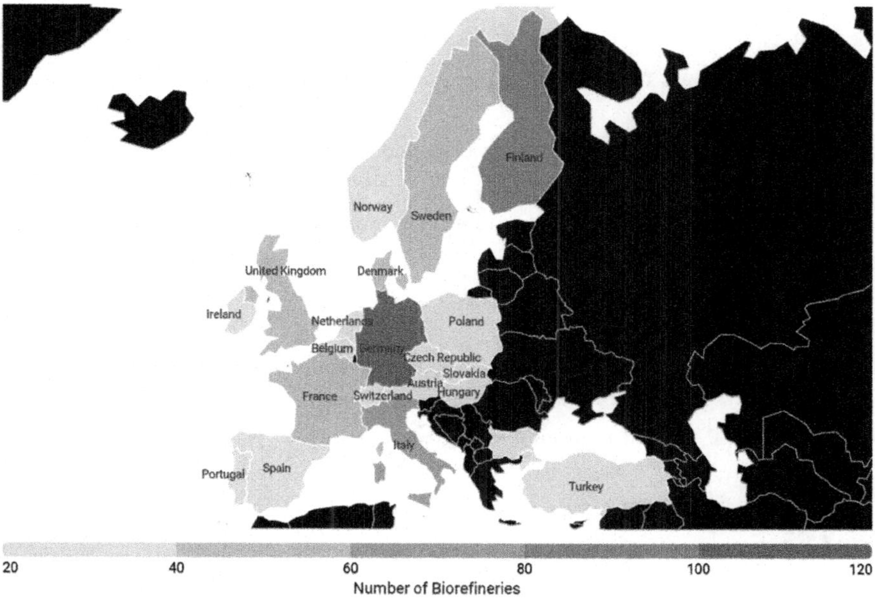

Fig. 3 Density map of European biorefineries

Switzerland were countries with less than 17 reported biorefineries. By all means, the list of reported biorefineries may not be completed to its fullest, however, it can be considered as a good approximation of the trends in Europe.

Among biorefinery studies, one of the most relevant concepts used to acknowledge the level of development of a process is the technology readiness level (TRL). This indicator can be used as a way of expressing maturity and development levels of technologies and innovative processes (Fig. 4). An approximated evaluation of the TRL of reported biorefineries has resulted in 82% of facilities with technology readiness levels of 8–9. Lower TRL of 6–7 and 4–5 were present in 9 and 8% of the cases respectively. The remaining 1% corresponds to biorefineries with technology readiness levels in the range of 1–3. With this, it can be stated that among reported biorefineries, most of them are considered to have technologies in operational environments that are considered to produce bioproducts and/or biofuels to a commercial level. The conclusion that can be drawn from this information is that biorefineries producing some kind of value-added product (in the form of materials or energy) in Europe are considered to have up-to-date or mature technologies. This in turn may signify that the development of bioprocesses in Europe is mainly centered towards processes that have been available for years now, rather than incurring in novel processes to produce specialty chemicals through innovative technologies. This would mean that Europe needs to take a step towards the development of researched processes in laboratory level and scale them to pilot or demo operations, to avoid stationary knowledge.

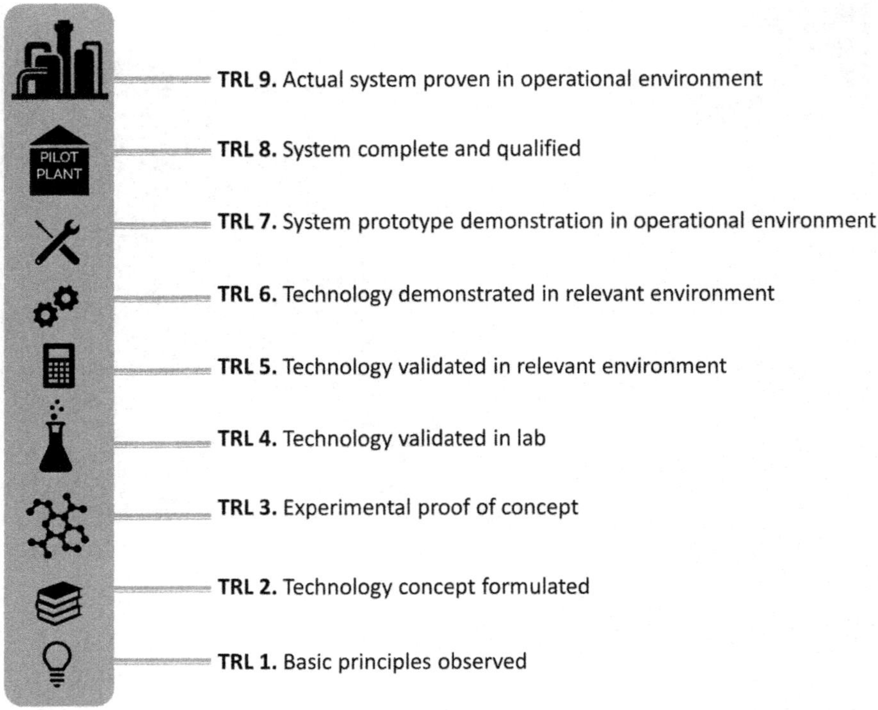

Fig. 4 Technology readiness level (TRL) diagram

Regarding the feedstock used in the considered biorefineries, the most frequent type of raw material was the category that englobes lignocellulosic materials. Around 38% of the reviewed facilities were englobed under the second generation biorefinery category. Under this group of feedstocks, the raw materials that could be frequently encountered along the different types of facilities were sawmill residues such as woodchips, bark or sawdust as well as residual crop fractions such as straw or grass. About 35% of facilities were plants that use blends of biomass with coal and peat. These were mostly power generation plants producing electricity and heat by means of burning biomass and coal. With less frequency, first generation biorefineries were identified (13%). Some of the most popular feedstocks within this category were wheat, corn, sugar beets and oil crops (rapeseed, sunflower, palm, soybean, kernel, coconut). Regarding fourth generation biorefineries, approximately 7% of processing plants were assigned to this group in the performed review. As defined in this study, fourth generation biorefineries are processing facilities that use as raw materials mostly residual fractions. Within the residual fractions available to be used as feedstock, some were sewage sludge, residual cooking oils, whey, manure and sulfite spent liquor streams from the pulp and paper industry. The remaining 7% of facilities were plants with possibilities to process different types of feedstock. These were usually combinations of raw materials from the different categories (first, second and fourth generation feedstocks). For instance, some facilities considered the use of blends of used oil, cooking oil and other residual fat and oil streams together with oil derived from oilseed crops. These biorefineries are usually intended to produce biodiesel and/or oleochemistry products such as fatty acids, glycerin, fatty alcohols, fatty amides, fatty esters, surfactants, methyl esters, paraffin waxes etc. It can be observed that biorefineries tend to use feedstocks that are commonly harvested or produced (as residues or as crops) nearby.

Very few times existing biorefineries were found to have the objective of producing fine chemicals such as furfural, levulinic acid, hydroxymethyl furfural or base chemicals as precursors of bio-based polymers or other chemicals. The reviewed plants are mostly producers of bioethanol, biodiesel, electricity and heat with little or no mention on the possibilities for recovery of side-stream bioproducts that would be feasible to be recovered alongside ethanol, sugar and pulp production processes.

4 The Energy Consumption Profile of a Biorefinery. LCB Case Study

Biorefineries have not being widely implemented at full scale, in fact, it is not yet possible to study issues such as energy integration on the basis of primary results. It is accurate to assert that the study of lignocellulosic biorefineries has been approached by literature in different ways.

The purpose of this study is to determine the impact of energy consumption in biorefinery facilities on environmental issues such as sustainable bioenergy production (in the form of bioethanol). The environmental study of a lignocellulosic biorefinery is presented as a case study to elucidate which would be the main impacts derived from the energy needs of a modelled facility with specific production characteristics. The intention of the LCA study is to provide an overview of the process areas and impact categories that are environmentally burdened by energy-related systems (energy consumption and energy production).

4.1 Materials and Methods

4.1.1 Goal and Scope

As mentioned before, LCA is a technique for assessing the potential environmental impacts associated with a product or process (ISO14044 2006a, b). The aim of this study is to assess the environmental performance of a lignocellulosic feedstock biorefinery by means of LCA methodology, considering the simulation work of

Kautto et al. (2013). The lignocellulosic biorefinery (LCB) produces high-added value products from residual woody biomass (waste stream from forest activities). The principles established by the ISO standards (ISO14040 2006; ISO14044 2006a, b) and the ILCD handbook (European Commission 2010) were followed in this research study. The functional unit considered for the assessment was the processing of 1 t/h of hardwood chips in the plant described in Sect. 4.1.2.

4.1.2 System Overview

Under the LCA approach, the analysis of a process should include defining clear boundaries and the processes within those boundaries that are required to be evaluated. In this study, the analysis of the biorefinery under assessment was carried out from a cradle to-gate perspective, considering the extraction of raw materials to produce the required products, but not the final disposal stage. All the activities involved, from the production of the raw materials to the final valorization processes of high-added value products in the biorefinery, were considered within the system boundaries, following the guidelines from other biorefinery works such as González-García et al. (2016), Laure et al. (2014) and Budzinski and Nitzsche (2016).

For the sake of clarity, the biorefinery was divided into subsystems (SS), which will be analyzed as independent blocks that compute to the total environmental impacts. The definition of different subsystems in the process makes it possible to identify the areas of the plant that represent a clear environmental burden for the entire system. Figure 5 shows a simplified block diagram of the production process and the identification of the main flows and subsystems.

The selected biomass to be exploited was hardwood chips, as a residual stream from a sawmill. The impacts associated to the raw material primary operations (SS0) in the sawmill were exclusively considered as the percent impacts directly related to the retrieval of the residual wood chips (González-García et al. 2014). This is made possible by implementing an economic allocation to the main product fractions in SS0 (bark, wood and wood chips).

Organosolv is a feasible pre-processing step to fraction lignocellulosic material. The organosolv process (SS1) of the system under study is based on the fractionation of beech wood at 180 °C for 60 min with ethanol and water (1:1 w/w) and 1.25% sulfuric acid as a catalyst. The delignification of wood through organosolv gives rise in two streams: liquor and pulp.

Subsystem 2 (SS2) includes all processing units that condition the liquor fraction and allow the recovery of the main non-energy based bioproducts (acetic acid, lignin and furfural). The liquor is mainly treated in a distillation column to recover the solvent (ethanol). The recovered ethanol is recycled back to SS1, resulting in a reduced fraction of the required fresh ethanol input. The reduction of the ethanol content favors lignin precipitation. Furfural is obtained as a co-product in a side stream of a distillation column that is still sent to a decanting unit (Kautto et al. 2013).

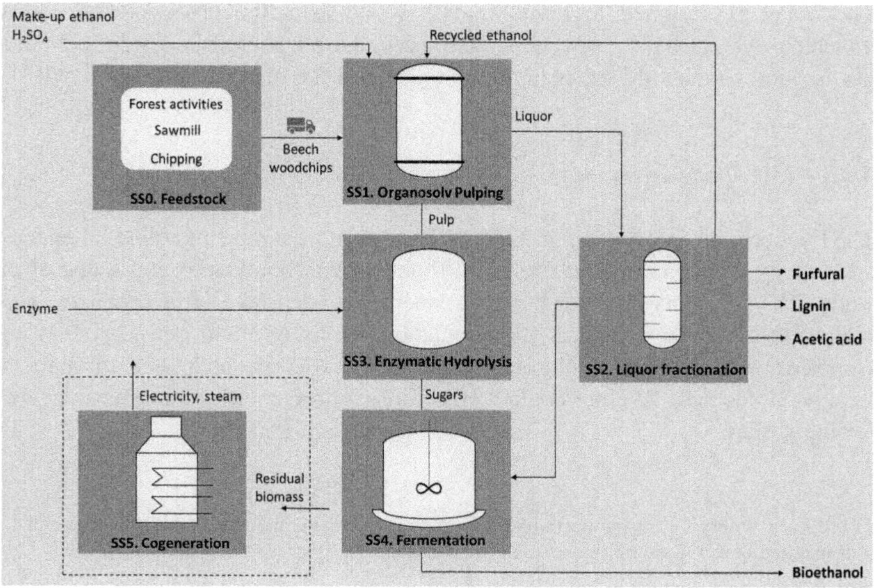

Fig. 5 Process diagram of the assessed wood-based biorefinery Adapted from Kautto et al. (2013)

After solvent recovery, water is evaporated from the liquor through a four-effect evaporation train. Evaporation allows low molecular weight lignin to be separated and burned as organic material in the boiler (SS5), along with other residual fractions. The resulting lignin-free hemicellulose fraction is used, together with glucose in the bioethanol fermentation (SS4). Acetic acid can be recovered from the condensates produced in the evaporation unit. Acid recovery is carried out through liquid-liquid extraction and distillation processes, with the use of trioctylphosphine oxide (TOPO) and undecane (Kautto et al. 2013).

The pulp or fiber fraction, washed with the ethanol-water solution before being discharged from the digester, is a cellulose rich fraction. The targeted objective for the pulp fraction is to transform the contained cellulose into second generation sugars, mainly glucose, which is performed in an enzymatic hydrolysis reactor (SS3), using an enzyme cocktail containing mainly cellulase. Cellulase production is considered within the system boundaries (Dunn et al. 2012; Heinzle et al. 2006) Cellulose hydrolysis is carried out at 48 °C for 84 h.

The resulting sugars in SS3 are further transformed in SS4, through fermentation by the microorganism *Zymomonas mobilis*. Bioethanol is recovered from SS4 with 99.9% purity by distillation and dehydration of the fermentation medium (Kautto et al. 2013).

The electricity and steam requirements of the biorefinery are covered through subsystem 5. This subsystem comprises the generation of the energy required to perform all transformation and valorization processes specifically considered in this case study. Subsystem 5 includes a boiler and a turbine for electricity production.

The boiler is designed to burn biomass waste from the process such as low molecular weight lignin, sewage sludge, bark and other organic residues. Natural gas is burnt to meet the energetic requirements of the plant (Kautto et al. 2013).

4.1.3 Life Cycle Inventory

The life cycle inventory stage in LCA is the collection of data regarding all material and energy inputs and outputs relevant to the system boundaries and scope of the study. The inventory data for this assessment has been taken from peer-reviewed bibliographic sources, such as the simulation results of Kautto et al. (2013) and González et al. (2014). Foreground processes have been addressed through the Ecoinvent database (Wernet et al. 2016). Table 3 presents the main inventory data for the system.

Table 3 Inventory for the lignocellulosic biorefining system defined for the functional unit of 1 t/h of residual beech wood chips including mass streams

SS0. Feedstock		
Overall inputs		
Diesel	1.03	kg
Fertilizer	0.15	kg
Water	244.38	kg
Chemicals	22.76	g
Packaging materials	0.27	kg
Lubricating oil	0.09	kg
Overall outputs		
Nitrogen emissions	52.68	g
Carbon emissions	30.77	kg
Emissions (SO_2)	0.82	g
Particulates	14.98	g
Municipal solid waste	1.50	kg
Residual woodchips	1.25	m^3
Bark	0.45	m^3
Sawn timber	1.8	m^3
SS1. Pulping		
Overall inputs		
Water	8.28	t
Sulfuric acid	1.01×10^{-2}	t
Ethanol	3.74	t
SS2. Liquor fractionation		
Overall inputs		
Water	1.02	t

(continued)

Table 3 (continued)

SS2. Liquor fractionation		
Ammonia	6.24×10^{-3}	t
Furfural (makeup)	1.92×10^{-3}	t
TOPO	3.32	kg
Undecane	11.93	kg
Overall outputs		
Acetic acid	1.56×10^{-2}	t
Furfural	5.28×10^{-3}	t
Lignin	0.16	t
SS3. Enzymatic hydrolysis		
Overall inputs		
Enzyme (cellulase)	7.80×10^{-3}	t
Cellulase production inputs (1 kg)		
Corn steep liquor	0.58	kg
Ammonia	7.82×10^{-2}	kg
Water	74.07	kg
Nutrients	0.32	kg
Cellulase production outputs		
N_2	0.28	g
O_2	0.84	g
CO_2	0.14	g
SS4. Fermentation		
Overall inputs		
Water	0.21	t
Corn steep liquor	1.27×10^{-2}	t
$(NH_4)_2HPO_4$	1.68×10^{-3}	t
Overall outputs		
Bioethanol	0.24	t
Water	0.37	t
CO_2	218.77	kg
O_2	1.44	kg
Wastewater	0.0034	m^3
SS5. Cogeneration		
Overall inputs		
Water	0.68	t
Sludge (WWT)	5.78×10^{-2}	t
Biogas (WWT)	4.27×10^{-2}	t
Natural gas	2.26×10^{-2}	t
Overall outputs		
CO_2	6.84	kg
Water (vapor)	108.84	kg

Table 4 Impact categories at midpoint level in the ReCiPe method analyzed in this study

Impact category	Acronym	Units of measure
Climate change	CC	kg CO_2 eq
Ozone depletion	OD	kg CFC-11 eq
Terrestrial acidification	TA	kg SO_2 eq
Photochemical oxidant formation	POF	kg NMVOC
Freshwater ecotoxicity	FET	kg 1,4-DB eq
Fossil depletion	FD	kg oil eq

4.1.4 Method

The environmental results were computed through the SimaPro 8.02 software by implementing the characterization factors from the ReCiPe 1.12 hierarchist method (Goedkoop et al. 2009). The evaluation of midpoint level impact categories was studied to determine the implications of the energy generation subsystem on LCA results. Although all categories of the ReCiPe method were studied, the environmental results are presented in terms of six impact categories relevant to the energy footprint in the system under study (Table 4).

The ReCiPe method has been popularly used in recent years on studies involving environmental assessment of biorefineries or biorefinery processes and includes impact categories that are considered to represent environmental characteristics of a system in a satisfactory manner (González-García et al. 2016; Lin et al. 2015; Parajuli et al. 2017; Silalertruksa et al. 2015). In this case study, selected categories have been targeted because of their relevance towards burdens from energy related activities.

4.1.5 Assumptions and Limitations

The limitations of the present study are mainly due to barriers on the data availability. For the studied system, mainly data from literature sources and databases has been utilized. Secondary data provides results with certain degree of uncertainty. In the same line, due to lack of available data, no infrastructure processes have been considered within the system.

4.2 Environmental Results and Discussion

The results of the environmental assessment indicate the process areas that cause an environmental burden. In the case of the energy footprint of this biorefinery, it is interesting to acknowledge subsystem 5, which is the cogeneration unit that

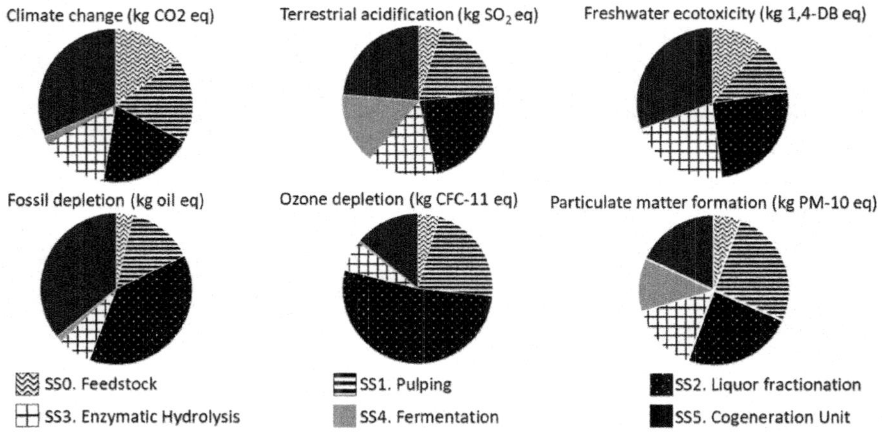

Fig. 6 Life cycle assessment results for six relevant impact categories depicting contributions originated on the process energy requirements in contrast with other processes in the biorefinery

provides energy in the form of electricity and steam to the entire production plant. The results for the six selected impact categories are presented in Fig. 6.

The results presented in Fig. 6 show that for most of the selected impact categories, the CHP unit in the biorefinery was responsible for a very representative percentage of the total impacts. Subsystem 5 contributed significantly to fossil resource depletion (36%), climate change (32%), land acidification (24%) and freshwater ecotoxicity (31%). The relevant contribution of SS5 to fossil depletion is due to the required input of natural gas to meet the energy demands of the process. On the other hand, SS5 is characterized by the handling of process residues to be burnt; therefore, it is important to highlight the importance of CO_2 emissions to air from this subsystem, which mainly contribute to the climate change category.

The energy consumption profile of the biorefinery can be described by two concepts: the impacts of SS5 due to the supply of energy and the energy required from the cogeneration unit. Firstly, the impacts associated with SS5 are the environmental impacts caused by the production of the energy required (Fig. 6). It includes, as mentioned above, the impacts of activities performed in subsystem 5, such as emissions from combustion and the use of natural gas. The overall impacts in SS5 account for, indirectly, all the subsystems in the biorefinery that consume energy. However, the determination of the overall contributions from SS5 does not provide an approximate idea of which areas of the plant represent the greater burdens on the energy footprint of the biorefinery. To determine the subsystems that are more burdening from an energetic point of view, Table 5 shows the energy consumption values of each process.

It can be clearly stated that that the most energy-intensive section of the process in an LCB is the pretreatment. This result agrees with the acknowledgment that other studies have performed on the economic relevance of the lignocellulosic biomass pretreatment, as viewed in Sect. 1.1.2.1. The energy consumption profile

Table 5 Electricity consumptions of the biorefinery for the exploitation of 1 t/h of woodchips Adapted from Kautto et al. (2013), González et al. (2014) and (Dunn et al. 2012)

Process section	Electricity consumption (kW)
Machinery use and chipping	3.6
Feed handling	30.0
Pretreatment	108.0
Cooking	30.0
Pulp washing	18.0
Reject screening	39.6
Pumps	9.6
Agitators	3.6
Compressors, screws, conveyors	7.2
Hydrolysis, fermentation and ethanol recovery	28.8
Pumps	6.0
Agitators	14.4
Compressors, screws, conveyors, mixers	7.2
Molecular sieves	1.2
Storage	0.1
Boiler and turbogenerator	19.2
Utilities	62.4
Cooling water pump and tower system	32.4
Chilled water system	22.8
Instrument air	2.4
Process water	1.2
Sterile water and CIP/CIS systems	3.6
Wastewater treatment	88.8
Enzyme production	1.3
Total energy consumption	342.2

of a biorefinery exploiting lignocellulosic biomass is defined in the pretreatment stage. In contrast to crop-based feedstock biorefineries, LCB pretreatment methods are based on more complex and energy-intensive processes to efficiently breakdown the feedstock and ensure efficient enzymatic hydrolysis (Tran et al. 2013; Zhu and Pan 2010).

4.2.1 Net Energy Gain

Many studies have defined and evaluated the net energy gain (NEG), also known as net energy value (NEV) in the production of biomass-based bioenergy (Arodudu et al. 2017; Illukpitiya et al. 2017; Luo et al. 2009). The NEG is a parameter that characterizes the net energy of a process producing biofuels such as bioethanol. This parameter can be obtained by subtracting the energy input (in the form of direct energy inputs from forestry and harvesting operations, processing and purification) from the energy output, in this case considered as the calorific value of ethanol. The NEG can be viewed as an effectiveness parameter to describe the biomass-to-fuel conversion process; it is a parameter that allows to compare different biorefineries producing ethanol. It can be considered as basis for the evaluation and achievement, for instance, of European targets for increasing energy from renewable sources (Arodudu et al. 2017). In fact, the production of energy from biomass resources can be considered sustainable when the net energy value is positive, therefore, when there is an accountable energy output (Zhu and Zhuang 2012).

As for this case study, there is an energy surplus considering the production of bioethanol, obtaining a positive NEG. Considering the lower heating value of ethanol as 26.8 MJ/kg, and the ethanol production per functional unit (Kautto et al. 2013), the total energy that can be released from ethanol is 1.76 MW per functional unit. Taking the total energy demand of the system under study per functional unit, as presented in Table 5, the estimated NEG for this process results in 17.1 MJ l^{-1}. The energy contained in biomass (beech wood) was not included as an input energy value in the calculation, only direct energy inputs were considered.

Different studies have presented the NEG parameter as a function of the input and output energy values. For instance, Illukpitiya et al. (2017) have determined the NEG for ethanol production from perennial grasses such as switchgrass, eastern gammagrass, big bluestem and indiangrass; on a volume basis, the calculated average net energy gain was 7.9, 5.8, 1.9 and 2.8 MJ·l^{-1} for each feedstock respectively. Farrell et al. (2006) assessed the bioethanol production from corn, and determined a net energy value in the range of 4–9 MJ l^{-1}. Schmer et al. (2008) have obtained a higher NEG value, with an average of 21.5 MJ l^{-1} for the production of cellulosic ethanol from switchgrass.

As can be seen, the NEG parameter presented positive values for different biorefinery systems; conclusively, it can be stated that the use of biomass for sustainable energy production presents positive results. Furthermore, it may be argued that grass species, lignocellulosic feedstock and energy crops show a high probability of resulting in positive NEG values (Illukpitiya et al. 2017).

4.3 Mapping the Environmental Impact of Electricity Consumption for Biorefineries

When analyzing the subject of biorefining under the approach of life cycle assessment, it can be observed that in many occasions, it is common practice to consider that most or part of the energy (in the form of electricity and heat) needed for the processes carried out in the facility is provided by a cogeneration unit available within the production scheme.

While this can mostly be realistic and common practice in many industrial clusters, there is a variable that is left out of the life cycle assessment scheme if energy is considered to be produced in a somewhat sustainable way within processing boundaries. This would be the potential need to use electricity from the grid rather than from a boiler or any similar cogeneration disposition.

On the other hand, it is usual to be generally familiar with the carbon footprint concept. Carbon footprint has been defined as an indicator of the overall balance of greenhouse gas emissions in a system expressed as CO_2 equivalent and based on a life cycle assessment approach (ISO 14067 2012). In the same lines, assessing the energy footprint of biorefineries may imply an evaluation of the energy consumed and or released to obtain a product or a service. However, in terms of energy, there is much more to energy footprint than energy consumption or production. In fact, one of the most relevant factors of energy production or consumption in terms of environmental sustainability is the origin of such energy. The use of fossil derived energy provides a non-sustainable source of electricity and heat for a process. However, if the electricity used in a process has been originated through renewable sources, its environmental burdens will most probably change notoriously.

Combining the fact that biorefineries may not be self-reliable energetically with the importance of energy provenance in energy footprint matters, the most natural step would be to evaluate how the electricity mix from the grid would affect sustainability when evaluated by means of LCA.

A hypothetical evaluation has been considered parting from the case study in Sect. 4.1. The system boundaries of the biorefinery case study considered the production of energy requirements through of subsystem 5. For the purpose of analyzing the impacts related to the source of energy, a sensitivity analysis was performed by means of eliminating subsystem 5 from the system boundaries of the biorefinery. Instead, it was considered that the electricity requirements for each subsystem were fulfilled through the use of electricity from the grid. Furthermore, it is known that the electricity mix in the grid is variant depending on the main energy generation methods of each country. Four countries with different electric mixes were considered for the present sensitivity study, in hopes of analyzing the environmental repercussions of the situational variable of a biorefining facility.

Spain, Turkey, Poland and Finland were the countries evaluated by means of the selection of their respective electricity mix process from the Ecoinvent v3.2 database. The comparative life cycle assessment results are presented in Fig. 7.

As it can be observed in Fig. 7, the differences in the hypothetical geographic location of a biorefinery are clearly influencing the overall environmental results of the impact assessment. Thus, the variation on the electricity country mix attending to location introduces the implications of the origin of energy production to the environmental results. Overall, Poland presents the worse environmental results in four of the evaluated impact categories (climate change, terrestrial acidification, freshwater ecotoxicity and fossil depletion). On the contrary, Finland presents the lowest environmental burdens in all the impact categories considered.

The electricity country mix in Turkey for the year 2015 was characterized by the predominance of natural gas and coal sources with 37.8 and 28.4% shares

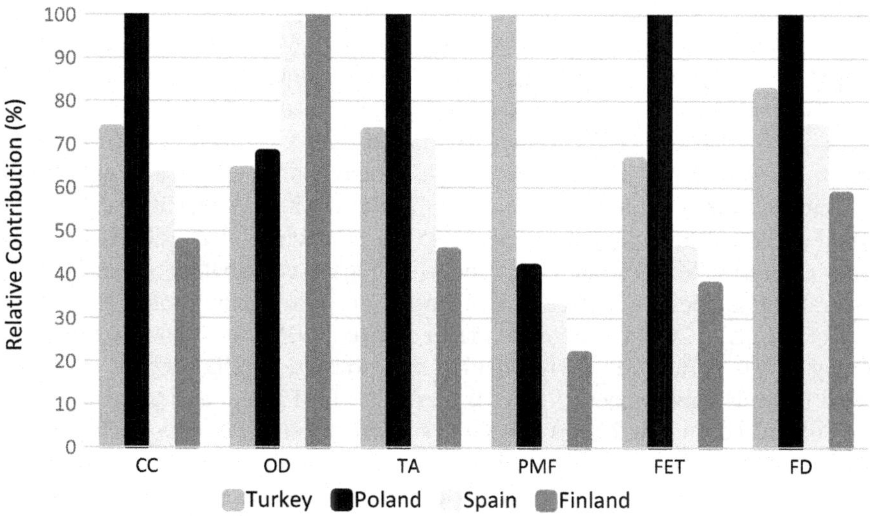

Fig. 7 Comparative analysis of the environmental profile of the biorefinery analyzing different electricity country mix from Turkey, Poland, Spain and Finland

respectively. Hydroelectric energy was produced in a percentage of 25.8%, while wind power represented 4.4% of the overall electricity produced in the country. Other electricity generation methods partake with lower percent values, such as geothermic energy with 1.3%, fuel, diesel and naphtha with a share of 1.6% and, finally, biogas with 0.6% (Ministry of Energy and Natural Resources. Republic of Turkey 2015).

For Poland, available data on the sources that participate in the overall electric mix is that of the year 2015. Coal and lignite power plants represented 71.6% contribution to the global electricity mix. Industrial power plants and gas fired utility power plants, on the other hand, took lower shares, with values of 6.2 and 3.2% respectively. Hydroelectric energy accounted for 5.9%, while wind and other renewable sources represented only 13.2% (Polskie Sieci Elektroenergetyczne (PSE) 2015).

In Spain, for the year 2017, 21.5% of the electricity demand was covered by nuclear power. Electricity produced through coal represented a share of 17% Combined cycle power plants produced electricity with a contribution of 13.9% Cogeneration plants represented 11% share. Electricity produced through the use of residual fractions (such as biogas) produced electricity with 1.2% shares. Regarding renewable sources, wind power represented 18.2%, hydroelectric power 7%, solar energy 5.2% and other renewable sources the remaining 1.4% of shares. Electricity imports for Spain represented 3.6% of overall contributions (Red Eléctrica de España 2017).

Finally, the profile of the electric mix in Finland for the year 2017 was composed by 25.2% of nuclear power, 23.9% imports, 12.8% biomass, 17.1% hydroelectric power. Other contributors to the electricity production in the country were natural gas (3.8%), coal (7.2%), oil (0.2%), wind power (5.6%), waste fuels (1.1%) and peat (3.1%) (Finnish Energy 2017). A summary table of the electricity supply contributions by source for each country is represented in Table 6.

As it can be observed through the data depicted in Table 6 as well as the environmental results presented in Fig. 7, the environmental impacts of a facility utilizing electricity from the grid are directly related with the percent contribution of fossil fuels and non-renewable sources to the production of the electricity in the mix. Countries like Turkey or Poland with higher contributions from coal and natural gas sources are clearly more impacted in environmental categories such as climate change (CC) or particulate matter formation (PMF). Contrarily to the tendency of CC, TA, PMF, FET and FD, ozone depletion (OD) depicts an inverse trend in which environmental impacts are higher for Finland and Spain and lower for Poland and Turkey. This is due to the significant contribution to electricity from nuclear power sources. Nuclear power is a carbon free energy source, however it uses uranium and therefore produces radioactive residues, as well as the emission of chlorofluorocarbons (CFCs), which are one of the most ozone depleting substances (Stamford and Azapagic 2011). The decarbonization of the electricity supply system would influence positively the environmental results of a facility. This will be beneficial towards energy security goals. From the results of the comparative evaluation (Fig. 7), it can be observed that the countries with higher percentage of renewable sources (wind power, solar energy, geothermic energy) have less environmentally burdening electricity mixes.

Table 6 Electricity supply contributions by source for Turkey, Poland, Spain and Finland in percent contributions (%)

Electricity source	Turkey[b]	Poland[b]	Spain[a]	Finland[a]
Coal	28.4	71.5	17	10.3
Natural gas	37.9	3.2	–	3.8
Fuel, diesel, naphtha, oil	1.6	–	–	0.2
Hydroelectric	25.8	5.9	7.0	17.1
Wind	4.4	–	18.2	5.6
Solar	–	–	5.2	–
Geothermic	1.3	–	–	–
Residues (Biogas, waste fuels…)	0.6	–	1.2	1.1
Nuclear	–	–	21.5	25.2
Industrial power plants	–	6.2	–	–
Combined cycle power plants	–	–	13.9	–
Cogeneration plants	–	–	11.0	–
Imports	–	–	3.6	23.9
Other renewable sources	–	13.2	1.4	12.8

[a]Data for the year 2017
[b]Data for the year 2015

5 Energy Footprint of Crude Oil Refineries

Biorefineries are developed under the assumption that they will lead to a comparably lower environmental footprint than oil refineries. The main idea on the production of sustainable chemicals lies on the basis that these facilities should environmentally outperform their petrochemical counterparts.

To the best of our knowledge, there are not many studies that directly compare the environmental impacts of the production processes in refineries and biorefineries. The overall sustainability advantages of biorefineries over oil refineries would hypothetically include reducing waste and emissions or increasing energy security by decreasing dependence on imported oil. However, crude oil refineries work with technology that has been developed for many years now and is implemented efficiently. The conclusion is that refineries and biorefineries are two very different systems, with different technologies, in very different situations and development levels that may not be comparable.

When considering a well to wheel environmental analysis, one of the main descriptors being evaluated is exhaust emissions from vehicles using the fuel produced, rather than exclusively considering the production process. It has been determined that vehicles using biodiesel (or other biofuels) produce less SO_2 and CO_2 emissions into the environment than conventional fuel (Bozbas 2008).

Energy for on-site use in oil refineries is generally derived entirely from fossil resources, in fact, from the actual crude oil feedstock. Therefore, it can be said that the consumption and production of energy in a refinery is directly related to the energy footprint of that facility, since fossil fuel is used both to produce energy and to be valorized in chemicals or products. Conventional refineries tend to have high energy efficiencies that result in positive net energy values, deriving in an export of electricity to the grid. The difference with biorefineries is the origin of the energy produced. Refineries produce 100% fossil-based energy, while the percentage of fossil origin in biorefineries tends to be reduced to a minimum. For instance, in the case study evaluated in Sect. 3, the contribution of natural gas to the boiler represents 3% of total contributions. Among different fuels produced in a refinery, energy use has been proven to be higher for gasoline, followed by diesel, LPG and naphtha. The same trend is followed by greenhouse gas emissions obtained from a well to pump (fueling station) perspective (Wang et al. 2004).

In refinery facilities, generally more than half of primary energy is used directly in the process, while the other half is distributed between steam generation and electricity generation. The energy footprint model for petroleum refining shows that energy use (electricity and steam) in manufacturing processes is largely dedicated to heating systems such as furnaces, reboilers or equipment using utility steam, followed by motor-driven units (pumps, fans, compressors). Process cooling and refrigeration as well as other energy-intensive activities represent lower shares. It has been demonstrated that, in general, the petrochemical sector is the largest consumer of fossil fuels among the manufacturing processes (Brueske et al. 2012).

6 Eco-efficiency and Energy Footprint in Biorefineries

Eco-efficiency has been proposed as an environmental management indicator to evaluate both the environmental impact and economic profitability of a process or a product (ISO 14045 2012). Some eco-efficiency indicators have been developed to report measurable quantitative values of sustainability in biorefineries. The relevance of the overall energy footprint of biorefinery facilities can be reflected in energy-related eco-efficiency indicators related to total biorefinery energy consumption, total energy consumption, non-renewable energy consumption or renewable energy consumption rate. They represent the energy consumed in the biorefinery allocated to the total benefit obtained from the products (Chua and Steinmüller 2010).

One of the main products that can be obtained from a biorefinery is electricity (as excess production). The assessment of eco-efficiency can therefore be made by assessing the economic benefit of selling electricity to the grid in relation to the environmental burden of its production. Silalertrusksa et al. (2015) have determined the eco-efficiency indicator defined as US$/kg CO_2 eq. for sugar cane refining products. The results showed that electricity was the most sustainable product, with the highest eco-efficiency values in all scenarios assessed.

The simultaneous consideration of environmental, energy and process parameters, together with the cost-effectiveness of the process in the design of a biorefinery is not easy. Usually, when applying fundamental concepts of life cycle assessment to processes or products, the procedure consists of proposing a series of recommendations for improving environmental parameters after having attained some results. An interesting perspective to guarantee eco-efficiency is the integration of Multi-Objective Optimization (MOO) in life cycle assessment; the result is a method that minimizes environmental impacts, considering specific constrains and objectives. A particular application to the present study would be to provide process alternatives for the optimization of eco-efficiency (environmental and economic factors) in biorefineries, providing, for example, sustainable alternatives for minimizing energy footprint (Mele et al. 2011; Vadenbo et al. 2017).

7 Energy Integration in Biorefineries

In describing the energy footprint of biorefineries, there is little published data on the energy consumption or energy efficiency of industrially implemented systems or real pilot scenarios. It is clear, when comparing the exploitation of crude oil and biomass valorizing methods, that the former is a mature process that has been improved over the last 100 years, while the latter is usually based on processes that still face numerous challenges. One of these challenges is the need for energy optimization.

Process optimization consists of implementing the minimum use of utilities, which would in fact result in minimal water and energy consumption. Optimization of the energy flows in the process implies the interconnection of these flows. This entails the utilization of residual heat in specific flows to provide heating power when necessary. For the optimal implementation of energy optimization methods, system connections should be kept to a minimum. Efficient optimization of energy consumption has the potential to increase the eco-efficiency of systems, for example, by extending the production catalogue without increasing energy input (Rafione et al. 2012).

It is clear that energetic integration has positive effects on reducing the environmental impacts of biorefineries by means of minimizing hot utility use and maximizing heat recovery (Celebi et al. 2017). However, energy integration may not only be achieved through the introduction of efficient interconnections within the same facility, but also through the establishment of global relationships at the industrial clusters. Research shows that the combination of supply systems from different plants in an industrial area presents significant opportunities to improve overall energy efficiency. Integration within industrial parks also makes renewable biomass exploitation opportunities accessible in existing units. Finding a market niche for biomass exploitation at the scale of oil refineries may be feasible as a result of integration into the current industry. The implications of such conclusions are directly related to the immediate possibility of reducing the ecological footprint (Hackl and Harvey 2013). A recognizable example is the implementation of the biorefinery concept for bioenergy production in the pulp and paper industry for an integrated approach (Bajpai 2013).

Beyond exclusive energy integration, some studies approach the feasibility of introducing on-site manufacturing of raw materials that have a major economic and environmental impact due to their production and transport. An example in an LCB is the integration of the enzyme production process necessary for enzymatic hydrolysis within the plant facilities. Enzyme production on-site indirectly decreases energy derived environmental impacts. Some studies suggest the possibility of avoiding enzyme concentration and purification operations after fermentation by means of using the whole culture broth and implementing simultaneous saccharification and fermentation (Olofsson et al. 2017).

8 Conclusions and Future Perspectives

Footprint studies aim to provide common ground on viable options for more sustainable development. In this case, the study of biorefining processes and their energy footprint makes it possible to establish a reference point to determine opportunities for future optimization and sustainable development.

To the best possible extent this chapter has aimed to analyze the biorefinery concept under the energy footprint approach. The objective was to present an overview of the most fundamental aspects of biomass exploitation under the life-cycle assessment approach. The energy footprint of biorefinery systems has been described by means of a bibliographic review that aims to cover the most recent developments, as well as by means of a case study, to specifically define relevant aspects on the energy footprint of biorefineries, such as net energy gain.

The results of this study show that the implementation of biorefineries in industry is a very reliable opportunity as a step towards meeting the sustainability and environmental objectives set out in European policies. The exploitation of biomass has a very favorable potential to replace or support the petrochemical industry. However, biorefineries have the capacity to become increasingly efficient through the implementation of optimization and energy integration methods. The energy footprint of biorefineries has a great capacity to be reduced to a minimum by applying the revised concepts. Future research should be geared towards the environmental study of the implications that energy integration matters can have. If fossil energy fuel in biorefineries is minimized, environmental burdens will tend to be diminished. However, it has been made clear that further comparative analysis among case studies contemplating optimization matters will provide key information on the steps to take regarding environmental sustainability of biorefineries.

Acknowledgements This research was supported by the Spanish Government (Ministry of Economy and Competitiveness) through the ERA-IB2 project 2G-Enzymes (PCIN-2015-031) and the European project STARProBio (Grant Agreement Number 727740). The authors belong to the Galician Competitive Research Group GRC ED431C 2017/29 and to the CRETUS Strategic Partnership (AGRUP2015/02). All these programmes are co-funded by FEDER (EU).

References

Adom FK, Dunn JB (2017) Life cycle analysis of corn-stover-derived polymer-grade l-lactic acid and ethyl lactate: greenhouse gas emissions and fossil energy consumption. Biofuels Bioprod Biorefin 11:258–268. https://doi.org/10.1002/bbb

Alvira P, Tomás-Pejó E, Ballesteros M, Negro MJ (2010) Pretreatment technologies for an efficient bioethanol production process based on enzymatic hydrolysis: a review. Bioresour Technol 101:4851–4861. https://doi.org/10.1016/j.biortech.2009.11.093

Arodudu OT, Helming K, Voinov A, Wiggering H (2017) Integrating agronomic factors into energy efficiency assessment of agro-bioenergy production—a case study of ethanol and biogas production from maize feedstock. Appl Energy 198:426–439. https://doi.org/10.1016/j.apenergy.2017.02.017

Aziz S, Sarkanen K (1989) Organosolv pulping—a review. Tappi J 72:169–175

Bajpai P (2013) Chapter 2—Biorefinery opportunities in the pulp and paper industry*. Biorefinery Pulp Pap Ind 11–15. https://doi.org/http://dx.doi.org/10.1016/B978-0-12-409508-3.00002-X

Barlow J, Sims RC, Quinn JC (2016) Techno-economic and life-cycle assessment of an attached growth algal biorefinery. Bioresour Technol 220:360–368. https://doi.org/10.1016/j.biortech.2016.08.091

Bioenarea (2010) Definition of the methods for the obtaining of active principles. Bioref

Biorefinery Euroview (2008) Workshop BioreFuture 2008 12/02/08 www.biorefinery-euroview.eu

Biorefinery Euroview, Biopol (2009) Joint deliverable report Biorefinery Euroview (addenda D1 . 2 and D1 . 3) & Biopol (D4 . 2) Note with results indentification, classification and mapping of existing EU biorefineries

Boisen A, Christensen TB, Fu W, Gorbanev YY, Hansen TS, Jensen JS, Klitgaard SK, Pedersen S, Riisager A, Ståhlberg T, Woodley JM (2009) Process integration for the conversion of glucose to 2,5-furandicarboxylic acid. Chem Eng Res Des 87:1318–1327

Bozbas K (2008) Biodiesel as an alternative motor fuel: production and policies in the European Union. Renew Sustain Energy Rev 12:542–552. https://doi.org/10.1016/j.rser.2005.06.001

Brueske S, Sabouni R, Zach C, Andres H (2012) U.S. Manufacturing energy use and greenhouse gas emissions analysis

Budzinski M, Nitzsche R (2016) Comparative economic and environmental assessment of four beech wood based biorefinery concepts. Bioresour Technol 216:613–621

Cassman K, Liska A (2007) Food and fuel for all: realistic or foolish? Biofuels Bioprod Biorefin 1:18–23. https://doi.org/10.1002/bbb

Celebi AD, Ensinas AV, Sharma S, Maréchal F (2017) Early-stage decision making approach for the selection of optimally integrated biorefinery processes. Energy 137:908–916. https://doi.org/10.1016/j.energy.2017.03.080

Chua CBH, Steinmüller H (2010) Development of eco-efficiency indicators for a biorefinery

Chum HL, Johnson DK, Black S, Baker J, Grohmann K, Sarkanen KV (1988) Organosolv pretreatment for enzymatic\rHydrolysis of poplars: I. Enzyme\rHydrolysis of cellulosic residues. Biotechnol Bioeng 31:643–649

Collet P, Lardon L, Hélias A, Bricout S, Lombaert-Valot I, Perrier B, Lépine O, Steyer JP, Bernard O (2014) Biodiesel from microalgae—life cycle assessment and recommendations for potential improvements. Renew Energy 71:525–533. https://doi.org/10.1016/j.renene.2014.06.009

Demirbas A (2010) Biorefineries: for biomass upgrading facilities. Green Energy Technol 26. https://doi.org/10.1007/978-1-84882-721-9

Dunn JB, Mueller S, Wang M, Han J (2012) Energy consumption and greenhouse gas emissions from enzyme and yeast manufacture for corn and cellulosic ethanol production. Biotechnol Lett 34:2259–2263

E4tech, RE-CORD, WUR (2015) From the sugar platform to biofuels and biochemicals. Final report for the European Commission, contract No. ENER/C2/423-2012/SI2.673791. https://doi.org/contract No. ENER/C2/423-2012/SI2.673791

Eaglesham A, Brown W, Hardy R (2000) The biobased economy of the twenty-first century: agriculture expanding into health, energy, chemicals, and materials. World, New York

Elvnert J (2009) Star-COLIBRI: strategic targets for 2020- collaboration initiative on biorefineries

Escamilla-Alvarado C, Poggi-Varaldo HM, Ponce-Noyola MT (2017) Bioenergy and bioproducts from municipal organic waste as alternative to landfilling: a comparative life cycle assessment with prospective application to Mexico. Environ Sci Pollut Res 24:25602–25617. https://doi.org/10.1007/s11356-016-6939-z

European Commission (2010) Energy 2020. A strategy for competitive, sustainable and secure energy. https://doi.org/COM(2010) 639

European Commission (2018) 2020 climate and energy package [WWW Document]. URL https://ec.europa.eu/clima/policies/strategies/2020_en. Accessed 15 Jan 18

Faraco V (2013) Lignocellulose conversion. Enzym Microbial Tools Bioethanol Prod. https://doi.org/10.1007/978-3-642-37861-4

Farrell AE, Plevin RJ, Turner BT, Jones AD, O'Hare M, Kammen DM (2006) Ethanol can contribute to energy and environmental goals. Science (80) 311:506–509

Farzad S, Mandegari MA, Guo M, Haigh KF, Shah N, Görgens JF (2017) Multi-product biorefineries from lignocelluloses: a pathway to revitalisation of the sugar industry? Biotechnol. Biofuels 10:1–24. https://doi.org/10.1186/s13068-017-0761-9

Finnish Energy (2017) Energy Year 2017. Electricity [WWW Document]. URL https://energia.fi/en

Fiorentino G, Ripa M, Mellino S, Fahd S, Ulgiati S (2014) Life cycle assessment of Brassica carinata biomass conversion to bioenergy and platform chemicals. J Clean Prod 66:174–187. https://doi.org/10.1016/j.jclepro.2013.11.043

Gavrilescu M (2014) Biorefinery systems: an overview. In: Bioenergy research: advances and applications, pp 219–241

Giwa A (2017) Comparative cradle-to-grave life cycle assessment of biogas production from marine algae and cattle manure biorefineries. Bioresour Technol 244:1470–1479. https://doi.org/10.1016/j.biortech.2017.05.143

Gnansounou E, Kenthorai Raman J (2016) Life cycle assessment of algae biodiesel and its co-products. Appl Energy 161:300–308. https://doi.org/10.1016/j.apenergy.2015.10.043

Gnansounou E, Pandey A (2017) Life-cycle assessment of biorefineries. Life-Cycle Assess Biorefin. https://doi.org/10.1016/B978-0-444-63585-3.00004-8

Goedkoop M, Heijungs R, Huijbregts M, De Schryver A, Struijs J, Zelm Van R (2009) ReCiPe 2008, A life cycle impact assessment method which comprises harmonised category indicators at the midpoint and the endpoint level. University of Leiden, Radboud University Nijmegen, RIVM, Bilthoven, Amersfoort, Netherlands

Gontia P, Janssen M (2016) Life cycle assessment of bio-based sodium polyacrylate production from pulp mill side streams: case study of thermo-mechanical and sulfite pulp mills. J Clean Prod 131:475–484. https://doi.org/10.1016/j.jclepro.2016.04.155

González-García S, Hospido A, Agnemo R, Svensson P, Selling E, Moreira MT, Feijoo G (2011) Environmental life cycle assessment of a Swedish dissolving pulp mill integrated biorefinery. J Ind Ecol 15:568–583. https://doi.org/10.1111/j.1530-9290.2011.00354.x

González-García S, Bonnesoeur V, Pizzi A, Feijoo G, Moreira MT (2014) Comparing environmental impacts of different forest management scenarios for maritime pine biomass production in France. J Clean Prod 64:356–367

González-García S, Gullón B, Rivas S, Feijoo G, Moreira MT (2016) Environmental performance of biomass refining into high-added value compounds. J Clean Prod 120:170–180

Hackl R, Harvey S (2013) Framework methodology for increased energy efficiency and renewable feedstock integration in industrial clusters. Appl Energy 112:1500–1509. https://doi.org/10.1016/j.apenergy.2013.03.083

Haddadi MH, Aiyelabegan HT, Negahdari B (2017) Advanced biotechnology in biorefinery: a new insight into municipal waste management to the production of high-value products. Int J Environ Sci Technol https://doi.org/10.1007/s13762-017-1424-x

Harmsen P, Huijgen W, López L, Bakker R (2010) Literature review of physical and chemical pretreatment processes for lignocellulosic biomass. Food Biobased Res, 1–49

Harris ZM, Spake R, Taylor G (2015) Land use change to bioenergy: a meta-analysis of soil carbon and GHG emissions. Biomass Bioenerg 82:27–39. https://doi.org/10.1016/j.biombioe.2015.05.008

Hatti-Kaul R (2010) Biorefineries- a path to sustainability? Crop Sci 50, S-152-S-156. https://doi.org/10.2135/cropsci2009.10.0563

Heinzle E, Biwer AP, Cooney CL (2006) Development of sustainable bioprocesses: modelling and assessment

Hernández V, Romero-García JM, Dávila JA, Castro E, Cardona CA (2014) Techno-economic and environmental assessment of an olive stone based biorefinery. Resour Conserv Recycl 92:145–150. https://doi.org/10.1016/j.resconrec.2014.09.008

Himmel ME, Ding S-Y, Johnson DK, Adney WS, Nimlos MR, Brady JW, Foust TD (2007) Biomass recalcitrance: engineering plants and enzymes for biofuels production. Science (80) 315:804–807. https://doi.org/10.1126/science.1137016

IEA Bioenergy (2018) Global database of biomass conversion facilities, including advanced biofuels, combustion, gasification and pyrolysis plants [WWW Document]. URL http://www.ieabioenergy.com/installations/

Illukpitiya P, Reddy KC, Bansal A (2017) Modeling net energy balance of ethanol production from native warm season grasses. Energy Econ 64:346–352. https://doi.org/10.1016/j.eneco.2017.04.008

ISO 14044 (2006a) ISO 14044: Life cycle assessment—requirements and guidelines. Int Organ Stand https://doi.org/10.1136/bmj.332.7550.1107

ISO14044 (2006b) ISO 14044, Environmental management—life cycle assessment—Requirements and guidelines 2006, 54

ISO 14045 (2012) ISO 14045:2012—Environmental management—eco-efficiency assessment of product systems–Principles, requirements and guidelines, 2012

ISO 14067 (2012) Carbon footprint of products—requirements and guidelines for quantification and communication

ISO14040 (2006) ISO 14040-Environmental management—life cycle assessment—principles and framework, ISO 14040. https://doi.org/10.1016/j.ecolind.2011.01.007

Kamm B, Kamm M (2004) Principles of biorefineries. Appl Microbiol Biotechnol 64:137–145. https://doi.org/10.1007/s00253-003-1537-7

Kamm B, Gruber PR (2006) Handbook of fuels beyond oil and gas : the methanol economy Bailey's industrial oil and fat products oil refineries in the 21st Century, biorefineries—industrial processes and products

Kamm B, Gruber PR, Kamm M (2008) Biorefineries-industrial processes and products: status quo and future directions, biorefineries-industrial processes and products: status quo and future directions. https://doi.org/10.1002/9783527619849

Kamm B, Gruber PR, Michael K (2016) Biorefineries—industrial processes and products. Ullmann's Encycl Ind Chem 1–38. https://doi.org/10.1002/14356007.l04_l01.pub2

Kautto J, Realff MJ, Ragauskas AJ (2013) Design and simulation of an organosolv process for bioethanol production. Biomass Convers Biorefin 3:199–212. https://doi.org/10.1007/s13399-013-0074-6

Khoshnevisan B, Rafiee S, Tabatabaei M, Ghanavati H, Mohtasebi SS, Rahimi V, Shafiei M, Angelidaki I, Karimi K (2017) Life cycle assessment of castor-based biorefinery: a well to wheel LCA. Int J Life Cycle Assess 1–18. https://doi.org/10.1007/s11367-017-1383-y

Kim S, Dale BE (2015) Comparing alternative cellulosic biomass biorefining systems: centralized versus distributed processing systems. Biomass Bioenerg 74:135–147. https://doi.org/10.1016/j.biombioe.2015.01.018

King D (2010) The future of industrial biorefineries. World Econ Forum Rep 1:1–38. https://doi.org/10.1086/294443

Kleinert M, Barth T (2008) Phenols from lignin. Chem Eng Technol 31:736–745. https://doi.org/10.1002/ceat.200800073

Laure S, Leschinsky M, Fröhling M, Schultmann F, Unkelbach G (2014) Assessment of an Organosolv lignocellulose biorefinery concept based on a material flow analysis of a pilot plant. Cellul Chem Technol 48:793–798

Levasseur A, Bahn O, Beloin-Saint-Pierre D, Marinova M, Vaillancourt K (2017) Assessing butanol from integrated forest biorefinery: a combined techno-economic and life cycle approach. Appl Energy 198:440–452. https://doi.org/10.1016/j.apenergy.2017.04.040

Li J, Wang Y, Yan B (2018) The hotspots of life cycle assessment for bioenergy: a review by social network analysis. Sci Total Environ 625:1301–1308. https://doi.org/10.1016/j.scitotenv.2018.01.030

Lin Z, Nikolakis V, Ierapetritou M (2015) Life cycle assessment of biobased p -xylene production. Ind Eng Chem Res 54:2366–2378. https://doi.org/10.1021/ie5037287

Luo L, van der Voet E, Huppes G (2009) An energy analysis of ethanol from cellulosic feedstock-Corn stover. Renew Sustain Energy Rev 13:2003–2011. https://doi.org/10.1016/j.rser.2009.01.016

Mandegari MA, Farzad S, van Rensburg E, Görgens JF (2017) Multi-criteria analysis of a biorefinery for co-production of lactic acid and ethanol from sugarcane lignocellulose. Biofuels Bioprod Biorefin 11:971–990. https://doi.org/10.1002/bbb

Martín M, Grossmann IE (2013) On the systematic synthesis of sustainable biorefineries. Ind Eng Chem Res 52:3044–3064. https://doi.org/10.1021/ie2030213

Mele FD, Kostin AM, Guillén-Gosálbez G, Jiménez L (2011) Multiobjective model for more sustainable fuel supply chains. A case study of the sugar cane industry in Argentina. Ind Eng Chem Res 50:4939–4958. https://doi.org/10.1021/ie101400g

Ministry of Energy and Natural Resources. Republic of Turkey (2015) Electricity mix 2015 [WWW Document]. URL http://www.enerji.gov.tr/en-US/Mainpage

Modahl IS, Brekke A, Valente C (2015) Environmental assessment of chemical products from a Norwegian biorefinery. J Clean Prod 94:247–259. https://doi.org/10.1016/j.jclepro.2015.01.054

Morgan-Sagastume F, Valentino F, Hjort M, Cirne D, Karabegovic L, Gerardin F, Johansson P, Karlsson A, Magnusson P, Alexandersson T, Bengtsson S, Majone M, Werker A (2014) Polyhydroxyalkanoate (PHA) production from sludge and municipal wastewater treatment. Water Sci Technol 69:177–184. https://doi.org/10.2166/wst.2013.643

Mosier N, Wyman C, Dale B, Elander R, Lee YY, Holtzapple M, Ladisch M (2005) Features of promising technologies for pretreatment of lignocellulosic biomass. Bioresour Technol 96:673–686. https://doi.org/10.1016/j.biortech.2004.06.025

Mosquera-Corral A, Carvalho G, Fra-Vázquez A, Ntaikou I, Oleskowicz-Popiel P, Palmiero-Sánchez T, Reis MAM, Suárez-Ojeda ME (2017) Recovery of organic added value products from wastewater. In: Innovative wastewater treatment & resource recovery technologies: impacts on energy, economy and environment, pp 399–421

Moussa HI, Elkamel A, Young SB (2016) Assessing energy performance of bio-based succinic acid production using LCA. J Clean Prod 139:761–769. https://doi.org/10.1016/j.jclepro.2016.08.104

Nova Institute, Consortium, B.B.I. (2017) Biorefineries in Europe 2017

Ofori-Boateng C, Lee KT (2014) An oil palm-based biorefinery concept for cellulosic ethanol and phytochemicals production: sustainability evaluation using exergetic life cycle assessment. Appl Therm Eng 62:90–104. https://doi.org/10.1016/j.applthermaleng.2013.09.022

Olofsson J, Barta Z, Börjesson P, Wallberg O (2017) Integrating enzyme fermentation in lignocellulosic ethanol production: life-cycle assessment and techno-economic analysis. Biotechnol Biofuels 10:51. https://doi.org/10.1186/s13068-017-0733-0

Pandey MP, Kim CS (2011) Lignin depolymerization and conversion: a review of thermochemical methods. Chem Eng Technol 34:29–41. https://doi.org/10.1002/ceat.201000270

Parajuli R, Knudsen MT, Birkved M, Djomo SN, Corona A, Dalgaard T (2017) Environmental impacts of producing bioethanol and biobased lactic acid from standalone and integrated biorefineries using a consequential and an attributional life cycle assessment approach. Sci Total Environ 598:497–512. https://doi.org/10.1016/j.scitotenv.2017.04.087

Pereira LG, Chagas MF, Dias MOS, Cavalett O, Bonomi A (2015) Life cycle assessment of butanol production in sugarcane biorefineries in Brazil. J Clean Prod 96:557–568. https://doi.org/10.1016/j.jclepro.2014.01.059

Polskie Sieci Elektroenergetyczne (PSE) (2015) Annual report 2015 [WWW Document]. URL https://www.pse.pl/web/pse-eng

Prasad A, Sotenko M, Blenkinsopp T, Coles SR (2016) Life cycle assessment of lignocellulosic biomass pretreatment methods in biofuel production. Int J Life Cycle Assess 21:44–50. https://doi.org/10.1007/s11367-015-0985-5

Rafione T, Marinova M, Montastruc L, Domenech S, Srinivasan B, Paris J (2012) Optimization of water and energy consumption in an integrated forest biorefinery. J Sci Technol For Prod Process 2:54–59

Rahimi V, Karimi K, Shafiei M, Naghavi R, Khoshnevisan B, Ghanavati H, Mohtasebi SS, Rafiee S, Tabatabaei M (2018) Well-to-wheel life cycle assessment of Eruca Sativa-based biorefinery. Renew Energy 117:135–149. https://doi.org/10.1016/j.renene.2017.10.035

Raman JK, Gnansounou E (2015) LCA of bioethanol and furfural production from vetiver. Bioresour Technol 185:202–210. https://doi.org/10.1016/j.biortech.2015.02.096

Red Eléctrica de España (2017) El sistema eléctrico español. Avance 2017 [WWW Document]. URL http://www.ree.es/es

Sarkar N, Ghosh SK, Bannerjee S, Aikat K (2012) Bioethanol production from agricultural wastes: an overview. Renew Energy 37:19–27. https://doi.org/10.1016/j.renene.2011.06.045

Schmer MR, Vogel KP, Mitchell RB, Perrin RK (2008) Net energy of cellulosic ethanol from switchgrass. Proc Natl Acad Sci 105:464–469. https://doi.org/10.1073/pnas.0704767105

Seghetta M, Hou X, Bastianoni S, Bjerre AB, Thomsen M (2016) Life cycle assessment of macroalgal biorefinery for the production of ethanol, proteins and fertilizers—a step towards a regenerative bioeconomy. J Clean Prod 137:1158–1169. https://doi.org/10.1016/j.jclepro.2016.07.195

Silalertruksa T, Gheewala SH, Pongpat P (2015) Sustainability assessment of sugarcane biorefinery and molasses ethanol production in Thailand using eco-efficiency indicator. Appl Energy 160:603–609. https://doi.org/10.1016/j.apenergy.2015.08.087

Souza SP, Gopal AR, Seabra JEA (2015) Life cycle assessment of biofuels from an integrated Brazilian algae-sugarcane biorefinery. Energy 81:373–381. https://doi.org/10.1016/j.energy.2014.12.050

Stamford L, Azapagic A (2011) Sustainability indicators for the assessment of nuclear power. Energy 36:6037–6057. https://doi.org/10.1016/j.energy.2011.08.011

Stuart P, El-Halwagi M (2014) Integrated biorefineries: design, analysis and optimization. https://doi.org/10.1017/CBO9781107415324.004

Tao L, Tan ECD, McCormick R, Zhang M, Aden A, He X, Zigler BT (2014) Techno-economic analysis and life-cycle assessment of cellulosic isobutanol and comparison with cellulosic ethanol and n-butanol. Biofuels Bioprod Biorefin 8:30–48. https://doi.org/10.1002/bbb

Tomei J, Helliwell R (2016) Food versus fuel? Going beyond biofuels. Land use policy 56:320–326. https://doi.org/10.1016/j.landusepol.2015.11.015

Tran UPN, Van Vu KL, Nguyen QD, Le PTK, Phan TD, Mochidzuki K, Kobayashi S, Seo D-J, Sakoda A (2013) Energy balance of small-scale biorefinery system. Environ Sci 26:489–496. https://doi.org/10.11353/sesj.26.489

Uihlein A, Schebek L (2009) Environmental impacts of a lignocellulose feedstock biorefinery system: an assessment. Biomass Bioenerg 33:793–802. https://doi.org/10.1016/j.biombioe.2008.12.001

Vadenbo C, Tonini D, Astrup TF (2017) Environmental multiobjective optimization of the use of biomass resources for energy. Environ Sci Technol 51:3575–3583. https://doi.org/10.1021/acs.est.6b06480

Vaskan P, Pachón ER, Gnansounou E (2018) Techno-economic and life-cycle assessments of biorefineries based on palm empty fruit bunches in Brazil. J Clean Prod 172:3655–3668. https://doi.org/10.1016/j.jclepro.2017.07.218

Vohra M, Manwar J, Manmode R, Padgilwar S, Patil S (2014) Bioethanol production: feedstock and current technologies. J Environ Chem Eng 2:573–584. https://doi.org/10.1016/j.jece.2013.10.013

Wang M, Lee H, Molburg J (2004) Allocation of energy use in petroleum refineries to petroleum products. Int J Life Cycle Assess 9:34–44. https://doi.org/10.1007/BF02978534

Wernet G, Bauer C, Steubing B, Reinhard J, Moreno-Ruiz E, Weidema B (2016) The ecoinvent database version 3 (part I): overview and methodology. Int J Life Cycle Assess 21:1218–1230. https://doi.org/10.1007/s11367-016-1087-8

Zheng Y, Pan Z, Zhang R (2009) Overview of biomass pretreatment for cellulosic ethanol production. Int J Agric Biol Eng 2:51–68. https://doi.org/10.3965/j.issn.1934-6344.2009.03.051-068

Zhu JY, Pan XJ (2010) Woody biomass pretreatment for cellulosic ethanol production: technology and energy consumption evaluation. Bioresour Technol 101:4992–5002. https://doi.org/10.1016/j.biortech.2009.11.007

Zhu JY, Zhuang XS (2012) Conceptual net energy output for biofuel production from lignocellulosic biomass through biorefining. Prog Energy Combust Sci 38:583–598. https://doi.org/10.1016/j.pecs.2012.03.007

A Technical Review on Methods and Tools for Evaluation of Energy Footprints, Impact on Buildings and Environment

Iheanacho H. Denwigwe, Olubayo M. Babatunde, Damilola E. Babatunde, Temitope J. Akintunde and Tolulope O. Akinbulire

Abstract The growing demand for clean and abundant energy in the society has a great impact on human well-being and biodiversity, studies on methods and tools for measurement of energy footprints are therefore necessary as energy footprints pose a barrier to clean energy in the society. This chapter presents knowledge and an understanding on energy footprints which is the measure of land required to absorb energy emissions, it focuses on the outcome of energy use by providing cases of energy emissions, analyzing tools and methods for measurements and finally highlighting problems of energy use to provide a guideline for corrective action to be taken. Different literatures for research studies on energy footprints measurement tools and methods are reviewed and discussed to provide an in-depth understanding on what energy footprint really is and its impact on buildings and the environment. It is concluded that having a proper understanding of the merits and demerits of different methods and tools for evaluation of energy footprints would aid in effective evaluation of energy footprints in buildings and the environment in general.

Keywords Energy footprints · Evaluation · Tools · Methods · Buildings Environment

I. H. Denwigwe (✉) · O. M. Babatunde · T. J. Akintunde · T. O. Akinbulire
Department of Electrical/Electronic Engineering, University of Lagos, Akoka, Yaba, Lagos, Nigeria
e-mail: Iheanachodenwigwe@gmail.com

D. E. Babatunde
Department of Chemical Engineering, Covenant University, Ota, Ogun State, Nigeria

S. S. Muthu (ed.), *Energy Footprints of the Bio-refinery, Hotel, and Building Sectors*, Environmental Footprints and Eco-design of Products and Processes,
https://doi.org/10.1007/978-981-13-2466-6_2

1 Introduction

Since the advent of the industrial revolution, climate change around the world has always been as a result of human activities and operations. Lots of existing scientific evidence show that rising temperatures and extreme weather conditions are caused largely by burning fossil fuels (Wright et al. 2011).

Presently, the need to urgently tackle the global warming threat has gained significant recognition around the world. In response to the growing threats posed by climate change, individuals involved in policymaking, government administration and running businesses have sought various ways and measures to ensure the reduction of emissions from greenhouse gases (GHG). The need to properly understand how human activities and operations bring about emissions from greenhouse gases and the search for credible measures to ensure the effective reduction of greenhouse gas emissions has made the idea of carbon footprinting for the estimation of GHG emissions popular (Wright et al. 2011).

The issue of climate change has garnered lots of attention around the globe. Scarcity of freshwater most importantly is a subject that has always been debated in government circles, the media and in business organizations. This is as a result of the impact which the scarcity of freshwater has on various economic sectors most especially in agriculture. The presence of freshwater in sufficient and adequate levels is a requirement for human societies, ecosystems and natural habitats. The gradual depletion of aquifers, drying of rivers and deterioration in the quality of water show that freshwater resources are not sustainably used by humans. This brings about a negative effect in societies as it endangers the health of ecosystems. There is, therefore, a need to identify sustainable solutions, approaches, and indicators for ensuring proper management and sustainable use of water for protection of the ecosystem while satisfying human needs of water and food.

The emission of Greenhouse gases (GHGs) occurs through many avenues which include clearance of goods on land, transportation, products and services, buildings, wood, roads, the production and consumption of food, fuels and manufactured goods (American Carbon Registry 2011).

Footprints can be obtained from direct sources like burning fuel from cars and stoves that are directly owned by individuals or from indirect sources such as fuel used by manufacturing plants for production of goods used by end consumers. According to a study by Jones and Kammen (2014), the majority of greenhouse gas emissions for households in the United States are obtained from indirect sources.

The phrase "carbon footprint" is developed from the concept of ecological footprints which is the estimation of the total area of earth surfaces required for all humans to consume an equal amount of resources. The use of Carbon footprints (energy footprint) is more specific and rampant as compared to the use of ecological footprints since carbon footprinting carries out direct measurement of greenhouse gas emissions which have negative impacts on the environment (Walther et al. 2005).

It is, therefore, very pertinent to explore means by which footprints of energy can be evaluated easily. Various methods and tools for achieving the purpose of evaluating greenhouse gas emissions exist globally. This chapter, therefore, carries out a technical review and analysis of these tools and methods. The latter part of this chapter is arranged as follows: Sect. 2 reviews various methods of evaluating energy footprints, Sect. 3 provides an overview of the impacts of energy evaluation methods, Sect. 4 provides a framework for developing an energy footprint evaluation model, Sect. 5 reviews the tools which can be used to evaluate energy footprints, Sect. 6 provides a framework for developing an energy footprint evaluation tool, Sect. 7 provides a framework for developing an energy footprint evaluation tool and Sect. 8 provides the conclusion of the chapter.

2 Methods of Evaluating Energy Footprint

The methods of evaluating energy footprints include Environmental Extended Input-Output Analysis (EEIO), Life-Cycle Assessment (LCA), Greenhouse Gas Accounting and Environmental Product Declaration

2.1 Environmental Extended Input-Output Analysis (EEIO)

Environmental-extended input-output (EEIO) analysis is a method of evaluation which provides a simple and effective method of checking carbon emissions by linking the contributions of industrial and economic activities to carbon emissions and the impacts on the environment (environmental impacts includes depletion and degrading of natural resources). EEIO analysis shows how various industrial sectors relate in the economy by depicting how the output from a particular sector of the economy could become the input to another sector of the economy. Summarily, EEIO can be used to evaluate energy footprints by assessing the impacts of direct (upstream) and indirect (downstream) sources of energy emissions; it can also be used to evaluate the impact of global trade of goods and services on the environment. The analysis carried out by Environmental-extended input-output (EEIO) method features input-output tables which account for emissions obtained in the course of carrying out economic activities (Ten Raa 2006; Miller and Blair 2009).

Kitzes (2013) gave a detailed preview of the goals, advantages and disadvantages of analysis on EEIO and also gave a detailed explanation on calculations, theories and principles involved in EEIO analysis. EEIO analysis thereby enables focus on consumption rather than production as a cause of global degradation of the environment and use of resources. Analysis involving the use of EEIO is very important as it can achieve some goals which include

- Evaluation of all impacts on the environment which could be direct (upstream), embodied and indirect (downstream). These impacts are to a large extent related to activities on human consumption of goods (indirect or downstream impacts), for example, the total emissions of carbon obtained from an individual's purchase and consumption of hamburger.
- Evaluation of the level of impacts embodied in global trade of goods and services between nations, for example, the release of nitrogen into environments in the United States after which it is consequently exportation in the form of wheat takes place via shipping to Denmark.

Analysis involving EEIO works by replacing the gas emissions obtained from the chain of production (processes or stages of production) with an input-output matrix which depicts monetary (financial) and sometimes physical flows of greenhouse gas emissions between various sectors of the economy, as an example, the amount of emission for carbon obtained in the agricultural sector of the economy through the production of poultry products can be estimated by measuring the carbon emitted from each produced poultry product sold to the public or by measuring the monetary gain which the sale of poultry products adds to the agricultural sector. This, therefore, provides an opportunity to track the impact of carbon emissions in trading goods and services between different sectors of the economy. It could also track the impact of emissions embodied in trade of goods and services between different nations through import and exports.

It is important to note that the input-output table sets the relationships between chains of production, carbon emissions and monetary gains from the sale of goods and services; this relationship is obtained or derived from known societal and environmental norms and conditions such as prices, accepted levels of emissions set by government guidelines and protocols. The table exists in the form of a matrix where the columns represent input to a particular economic sector while the rows represent output from a particular economic sector thereby showing the dependency between various economic sectors by depicting the monetary value of both inputs and outputs. The input-output features variables such as economic activities or sectors, total inputs, total outputs, value added, final demand, exports etc.

According to Kitzes (2013), EEIO analysis presents advantages such as

- review or follow-up of chains or stages of production to check for performance or for verification of obtained results;
- provision of means for making proper inference on the most acceptable techniques and methods of producing goods and services through the use of input-output tables;
- addressing loops or cycles which exists in practices of production or which can be found in products or materials used directly in the process of production;
- avoiding mutually exclusive allocation of impacts between economic activities and sectors which leads to double counting;

- proper capturing of the trade involving goods, products and services especially with the use of a monetary input-output table for proper evaluation of impacts on the environment through emission of greenhouse gases.

Some of the disadvantages of EEIO analysis (Kitzes 2013) include

- Difficulty in getting the true state of impacts by emissions from different economic sectors through the assumption that trade of goods and services between different sectors is homogenous in terms of prices, outputs and emissions bring limitations when carrying EEIO analysis. For instance, two different goods from different economic sectors sold at the same price in a particular economic sector bring up the assumption that the goods are the same or the goods are identical in terms of impacts on the environment obtained from emissions.
- The number of economic sectors or activities which can be added to the input-output table is sometimes low making it difficult to track environmental impacts related to commodities related to sectors which are not in the input-output tables.
- The inability of input-output tables to capture all economic activities has a negative effect in low-income nations where there are no records on environmental impacts of activities.
- The differences and inequality in the computation and collation of raw data in different countries limits the accuracy of input-output tables used globally.
- Environmental impact inventories of nations often depict data and modelled estimates which are characterized by bias and uncertainties, this, therefore, affects the accuracy of environmental impact assessments and evaluation.

Tukker et al. (2009) developed an EEIO framework which estimates the impact of energy footprints or external costs from the economic sector and consumption activities on the environment. According to the European System of Accounts (ESA95), the framework of input-output analysis consists of supply and use tables (SUT) (which shows the supply and use of goods and services) and symmetric Input-Output Tables (IOT) (European Commission, IMF, OECD, and United Nations 1993; EC and OECD 2008), an IOT can be derived from an SUT using various technological assumptions and standards. According to Tukker et al. (2009), the assumption by analysts that the technology used in the production of imported goods is the same with the technology used in the production of domestic goods can result in great errors can be resolved using multi-regional Input-output models through the use of IOTs, identification of trading partners of a particular country, calculation of resources used in trade between two countries and calculation of pollution embodied in trade between two countries. Tukker et al. (2009) identified some problems which limit the effectiveness of the data used in EEIO for the assessment of environmental impact via domestic and foreign trade activities of a country. They include (i) the provision of Input-Output Tables for countries having no trade links with other countries; (ii) Poor information on economic sectors, produced goods and services in a country; (iii) Absence of information on use of resources and emission types; (iv) Data across different countries are not properly harmonized. An EEIO database is very crucial in the development of environmental and economic policies by improving and monitoring of evaluation of issues on environmental impact and external costs by emissions from sectors of industries, consumption activities, imports and exports. Analysis of life-cycle impacts, external costs and life style patterns is made possible with EEIO through the breakdown of consumption activities into patterns of consumption for different target groups makes EEIO a very good method for energy footprint analysis (Stadler et al. 2014).

Turner et al. (2007) Explored the use of EEIO in measuring the use of resources, generation of pollution and ecological footprints which are embodied in trading activities. They defined EEIO analysis as a set of economic activities which are disaggregated in terms of sectors where inputs and outputs of the sectors are separately identified. The primary function of EEIO in relation to energy footprints is to show how the interdependence of different economic activities leads to the emission of greenhouse gases with the use of Input-output tables which are designed in units of monetary representations for purposes of national accounting. According to Miller and Blair (2009), the central input equation for EEIO analysis is

$$\mathrm{a} = (H - K)^{-1} b \tag{1}$$

where

- 'a' represents gross output N × 1 vectors with elements a_i where i = 1, …, N for each sector of the economy
- 'b' represents final demands N × 1 vectors with elements b_i where i = 1, …, N for each sector of the economy
- 'K' represents a matrix of the requirements for EEIO analysis or coefficients of input-output
- $(H - K)^{-1}$ represents the Leontief inverse which accounts for the output generated in one sector which is as a result of the final demand in another sector.

The full use of resources or generation of pollution/emissions can be determined using

$$g^a = \emptyset a \tag{2}$$

where

- g^a is a vector with elements g_k^a (where k = 1, …, K) which represents resource use k during economic activities of production
- Ø is a matrix which shows how the use of resources interact with gross output from economic sectors.

By combining Eqs. 1 and 2, we obtain the standard input-output equation for EEIO analysis as

$$g^b = \varnothing(H - K)^{-1}b \tag{3}$$

where

- g^b is a vector with elements g_k^b (where k = 1, …, K) which represents requirements of direct or indirect resource use, k, needed to meet total final demand, b, in the economic activities of production.

The application of input-output method for the evaluation of energy or ecological footprints can be done by populating the coefficients of the matrix, Ø, with coefficients of energy or ecological footprints, this can be done by breaking down the records of energy or ecological footprints associated with production in a country. After the breakdown, Eq. 3 can then be used to determine the different types of consumption activities that directly and indirectly bring about the pre-existing records of energy footprint. In some other applications of energy footprints evaluation, metrics such as land-use coefficients were used to populate the matrix, Ø, (Bicknell et al. 1998; McGregor et al. 2008; Lenzen and Murray 2001).

2.2 Life-Cycle Assessment (LCA)

Life-cycle assessment (LCA) is a systematic method for analysis and assessment of impacts occurring in the environment over the course of the full life-cycle period of a product or activity. There are two types of life-cycle assessment namely: Process-based life-cycle assessment and Economic input-output life-cycle assessment (EIO-LCA) (Simonen 2014).

2.2.1 Process-Based Life-Cycle Assessment (LCA)

Process-based Life-cycle assessment is an approach used in the implementation of the life-cycle assessment method which carries out assessments on the impacts of emissions on the environment which is associated with the product's life stages beginning from extraction of raw material through stages of processing materials, manufacturing, distributing, use of material, repair and maintenance of material, and finally recycling or disposing the material. Life-cycle assessments provide databases on emissions, materials capable of releasing emissions to the environment, evaluation of environmental impacts from emissions and interpretation of evaluation results to make decisions which positively affect climate change policy to solve environmental issues and concerns.

Life-cycle assessments provide a comparison between environmental effects triggered by the use of products and services by determining the extent to which material flows of inputs and outputs affect the environment. LCA consists of a

Life-cycle Inventory (LCI) which is the database or data collection platform of the LCA. The Life-cycle Inventory is where information on monitored and assessed emissions is stored (UNEP/SETAC Life cycle Initiative 2005).

Khasreen et al. (2009) reviewed life-cycle analysis as a decision-making tool for evaluation of environmental impacts of carbon emissions from buildings by assessing the environmental loads of activities and operations during the entire period of the building life-cycle. The methodology of life-cycle assessment consists of four different steps which include:

i. **Goal and Scope definition**: The goal and scope definition involves an important step of identifying the purpose for which the study was established and determining the questions needed to be answered to meet the purpose of the study. This step is responsible for the formation of all objectives, constraints and limitations of the study thereby setting vital assumptions such as system boundaries identification, phase of production (e.g. data quality, functional unit such as floor area etc.) and product's full life time. Ultimately, the exercise of goal and scope definition defines important benchmarks and provides the direction which the study is to follow. Over the course of carrying out the study, there is likelihood for a change to occur in the goal and scope of study, this could be as a result of factors like unavailability of data and insignificance of impact. ISO 14040, the international standard for life-cycle assessment states that the characteristics of the goal of a life-cycle assessment should include the planned area of application, reasons for pursuing the work of study and the targeted audience (which consists of data requirements, procedures of allocation, functions, selected categories of impact, functional units, boundaries, selected categories of impact, impact assessment methods, limitations, assumptions, requirements on initial data quality, report required for study, type of critical review and systems of the product to be studied). Generally, the goal of carrying out Life-cycle assessments in buildings is to reduce the negative impact which the building has on the environment over the entire period of the building's life-cycle (Rebitzer et al. 2004).

ii. Creation of Life-cycle inventory or Inventory analysis: The step of inventory analysis or life-cycle inventory involves every aspect of collecting data and all procedures involved in calculation, this is considered a very important step or stage in Life-cycle analysis as data is important for analysis which forms the basis of the study. There is a close relationship between the life-cycle inventory and definition of scope as collection of data and other inventory issues (e.g. lack of data) may lead to the reformation or redefinition of the system boundaries involved in the scope definition. According to ISO 14040, life-cycle inventory analysis involves collection of data from high-quality materials and resources that are readily accessible and available, validation of obtained data, establishing the relationship between data and unit processes, establishing the relationship between calculated data, functional units and procedures of allocation in systems of multiple products and recycling platforms. It is important to note, that the more the quality of obtained data, the more quality the result of the analysis is. The success or failure of the study is, therefore, dependent on the quality of data, the origin or source of data is, therefore, very important when considering and choosing data needed for analysis. Assessment of data quality depends on completeness, reliability, and technological, temporal and geographical correlation. Collection of data involves input-output energy data, mass flow of carbon emissions to water, air and land. The phase of Life-cycle Inventory makes use of databases of emissions obtained from combination of components and building materials (Rebitzer et al. 2004; Madrazo et al. 2018).

iii. Life-cycle Impact assessment: Life-cycle assessment is very useful in the building industry as it helps in the identification of sustainable practices for building construction. The application of Life-cycle assessment to a building could involve the definition of assessment according to a certain level called whole process of a building which contains all material processes, the application could also involve the definition of assessment according to a certain level called Building Material and Component Combination (BMCC) which deals with assessment of a building part, building material or component of a building where the recognition of the component impact equivalent according to the building's functional unit is very important. Life-cycle impact assessment (LCIA) is a framework involving multiple procedures and steps which include.

- Selection and definition of relevant categories of impact such as acidification, toxicity and global warming;
- Classification of results obtained from the phase of Life-cycle Inventory according to different categories of impact (e.g. classification of emissions of CO_2 *as a cause of global warming*);
- Comparing different impact to state a potential or higher impact e.g. comparing the impact of global warming obtained from the emissions of methane and carbon dioxide respectively;
- Evaluation of results from comparison, grouping of potential impact according to their level or height effect (for instance, global warming has an effect which is felt around the globe) and reporting the results and all procedures and operations involving Life-cycle Impact Analysis.
- ISO 14042 describes Life-cycle impact assessment as the employment of categories of impacts and indicators which are obtained from the phase of Life-cycle Inventory for the examination of systems and products from the standpoint of impacts on the environment. The categories of impact could differ as a result of factors such as availability of data, goal of study and impact significance.

iv. Result interpretation: Interpretation of results involves analyzing the obtained results from the impact assessment, making inferences, giving proper explanations to the encountered limitations and recommending measures and proposals which are linked to the knowledge gained in the inventory analysis and impact assessment phase.

2.2.2 Economic Input-Output Life-Cycle Assessment (EIO-LCA)

Economic Input-output life-cycle assessment is an approach used for the implementation of the Life-cycle assessment method. It provides an estimate of the emissions obtained from energy materials and resources used for carrying out economic activities. EIO-LCA is a combination of input-output analysis and life-cycle assessment. It is a Life-cycle assessment technique which uses information on economic activities to estimate emissions obtained from production and supply chains throughout the period of the economic life-cycle. It is a mathematical procedure which involves the use of data on environmental and economic data for the determination of the effect obtained from the change encountered in one economic sector. Mathematically, it occurs as a matrix which shows how economic activities between different industrial sectors constitute emissions to the environment. In the form of a model matrix, the columns represent different sections of the environment (land, sea and air) while the rows represent emissions as an output from an economic sector which is input to the environment sections. The inclusion of emissions and interactions between economic sectors in EIO-LCA makes system boundaries needed for the assessment broad and clear thereby eliminating the need for the step of scope definition required in process-based life-cycle assessment.

According to Joshi (1999), EIO-LCA can only compare different components from different economic sectors but is unable to compare different components in a particular economic sector, this is due to the approximation of products by the product's economic sector in the national input-table which uses input requirements and environmental coefficients in approximation, this thereby limits the success of the assessment. Another limitation to EIO-LCA of environmental impact is that EIO-LCA only records the negative upstream environmental effects obtained from the acquisition of raw materials and stages of production but fails to record effects obtained from the use of product and end-of-life options like disposal or recycling. Joshi (1999) provided six options for the implementation of EIO-LCA, they include

1. Economic sector approximation of the input and output of the product or service: The economic sector approximates the input and output products and services with respect to coefficients of environmental and technical burden. The effect occurring directly or indirectly by incrementing the output of a product or service can be estimated by handling it as an external change in demand for the output of the sector.
2. Presenting product or services as an assumed or proposed (hypothetical) economic sector: Products of services can be presented as a new sector being introduced to the economy as a hypothesis if direct burdens of environment from the process of production and the information regarding inputs needed for the production of products are available.
3. Breaking down an already existing sector of the economy: This involves the breakdown of an economic sector into two or more sectors when the total burden on the environment related to changes in external demand of a product which is included in one of the already existing sectors needs estimation.

4. Repeated breakdown of economic sectors when the availability of a conventional LCA which is limited in use is present: This involves the development of an expansion of coefficients of technical and environmental burden by repeatedly breaking down all functioning and relevant sectors.
5. Including the product's use phase of the product's life-cycle: This involves the treatment of the product's use phase as a hypothetical sector of the economy which takes input related to some environmental burdens from existing sectors of the economy in EIO-LCA. The related environmental burden can be estimated by the impact which the output of the product's use phase which is treated as a hypothetical economic sector has on the economy. The assumption here is that the economic matrix of technical coefficients is constant throughout the product's lifetime and the time rate of discount for burdens on the environment is zero.
6. Including Management options for end-of-life stage: This involves the treatment of end-of-life management options such as remanufacturing, reuse, recycling or disposal as an extra hypothetical economic sector in the analysis involving EIO-LCA.

Table 1 compares the advantages and disadvantages of EIO-LCA with process-based LCA

Table 1 Advantages and disadvantages of EIO-LCA and process-based LCA

	EIO-LCA	Process-based LCA
Advantages	(1) It makes use of results which are available in public and is therefore capable of converting those results (if there are lapses) into different forms for further use	(1) It recognizes areas where processes need to be improved and allows for analysis of weak points in results
	(2) Results obtained are very comprehensive in assessment of economic sectors	(2) Obtained results are very detailed and specific in assessing processes
	(3) It allows comparison of different levels of economic systems or sectors	(3) It allows comparison of different processes and products
	(4) It makes provision for assessment of future development of economic sectors. It gives information on all economic commodities	(4) It makes provision for assessment of future development of products and processes
Disadvantages	(1) It is very difficult to carry out assessments	(1) It is costly and takes a lot of time
	(2) Results are limited as monetary values have to be linked to physical units	(2) Application to new design processes is difficult
	(3) Data must be available for the method to have complete effect on the environment	(3) If private data is used in a particular process, the method cannot be used in another process

2.3 Greenhouse Gas Accounting

Greenhouse gas accounting is a method of measuring energy footprints which involves auditing and keeping a database on emissions of greenhouse gases. It is a systematic assessment of the direct and indirect production of all greenhouse gases obtained from business activities and operations. Generally, it involves the processes of monitoring, reporting and inventorying. Greenhouse gas accounting exists in two forms namely: consumption-based accounting and production-based accounting

1. Production-based accounting: Production-based emissions accounting involves a well-established method which indirectly calculates emissions obtained from the use of greenhouse gases and other relevant processes and activities of production in the economy. The high point of this method is the presence of a well-established methodology which assists in policy formulation and implementation while the inability to account for emissions embodied in imports limits the performance of the method (Liu et al. 2015; Boitier 2012).
2. Consumption-based accounting: Consumption-based emissions uses Input-Output Tables to depict the relationship and interdependence between different economic sectors in terms of production activities and provides features for tracing the effects of production and consumption activities in a nation's economy. The steady increase in the development of international trade in terms of import and export activities has brought about the development of consumption-based multi-regional input-output (MRIO) greenhouse gas accounting models. Multi-regional input-output (MRIO) greenhouse gas accounting models uniquely allows the tracing of effects of production activities throughout the period of their life-cycle and also highlights how consumed products and services from different economic sectors contribute to emissions released to the environments, thereby describing the processes involved in the global supply chains of consumed products. The availability of economic data on carbon emissions for each output allows for the evaluation of the total amount of emissions from the product and the final consumer of the product (Liu et al. 2015; Boitier 2012). The merits of consumption-based accounting include incorporation of embodied emissions uncovered by production-based accounting, giving room for increased participation to cover more global emissions, extension of options for mitigation (through the use of methods like harmonized global tax on emissions and border-tax adjustment on emissions) and the use of policies like Clean Development Mechanism (CDM) to encourage environmental-friendly behaviors and to promote technologies for clean production. The demerits of consumption-based accounting include increased uncertainty and complexity due to unavailability and poor quality (inaccuracy) of data, and the success of consumption-based accounting depends on high-level cooperation between countries, for example, if countries oppose internationally agreed conventions on importation habits and methods of production, the concept of consumption-based accounting will be limited in action.

The two greenhouse accounting methodologies possess certain differences and similarities. While Production-based accounting takes the evaluation of the monetary value of produced goods and services into consideration, consumption-based accounting which has more levels of uncertainty and complexity takes activities of trade and national consumption. Consumption-based accounting notably considers cases global emissions especially emissions are not accounted by production-based accounting e.g. emissions which are released throughout the life-cycle of a product i.e. emissions released after stage of product creation to stage of disposal. Consumption-based consumption provides global options for mitigating the release of emissions. Factors like mitigation options and accounting for product life-cycle emissions make consumption-based accounting more preferable when compared to production-based accounting (Schaltegger and Csutora 2012).

2.4 Environmental Product Declaration (EPD)

Environmental product declaration is a method of energy footprint used for the communication of a product's environmental performance which is then used for decision-making and policy proposals for climate change. The method of environmental performance declaration exists as schemes and programs conducted by organizations and corporations. Environmental performance declaration schemes make use of many environmental indicators used for assessing, monitoring and reporting the environmental performance of products and services. Environmental product declaration is also used for reporting the results of life-cycle assessments by providing standard categories of information sets using the product category rule (which outlines the rules and requirements for environmental declarations of products in specific product categories) and the ISO 14025 format of reporting. With Product Category Rule (PCR), EPD can be used to directly compare products.

Recent studies on Environmental Product declaration has shown the need for comparability, harmonization and streamlining of the EPD process to be taken seriously when developing EPD schemes. According to Ingwersen and Stevenson (2012), comparing the environmental performance of different products and services leads to the prevention of falsified claims regarding product declarations and environmental impacts. Borghi (2013) stressed the need for EPD schemes and programs to be harmonized to ensure that product declarations are comparable and transparent, and also to ensure that relevant rules are not duplicated. Streamlining of the EPD process widens the application of EPD, Zackrisson et al. (2008) presented method of streamlining known as stepwise EPD and the examination of data assistant tools for streamlining by Fet et al. (2009) shows how the streamlining of the EPD process assists small and medium enterprises that have their resources and expertise limited in supply.

The four steps of carrying out environmental product declaration include (Manzini et al. 2006)

1. Development of Product Category Rule (PCR) or use of an already existing PCR: PCRs can be developed through partnership between companies and organizations after consultations with experts in the field of Life-cycle Assessments (LCA). The PCR consists of rules, requirements, protocols and reporting standards.
2. Carrying out Life-cycle assessments: This involves assessing the product's composition, extraction and manufacturing processes of the product as well as resource use and the impact on upstream and downstream sectors of the environment.
3. Development, verification, and publishing of EPDs: This involves the establishment of EPD scheme after the life-cycle assessment has been completed. This is done using established PCR tools or model.

It is also very important to properly understand EPD schemes and programs, to ensure a successful evaluation of energy footprints in the environment. According to Borgeskar et al. (2002), the guidelines that need to be followed when reading and understanding schemes of EPD include

1. Making sure that the EPD scheme is certified by a trusted external source e.g. the program operator.
2. Making sure that the period of certification has not expired.
3. Making sure that all products to be compared are using the same PCR format.
4. Comparing the categories of life-cycle impact for the product with the environmental objectives chosen for that particular product. This is done for easy simplification of products characteristics and impacts.
5. Begin the reading with the portion of the EPD which summarizes key results obtained from the EPD scheme.

Tasaki et al. (2017) organized a survey for EPD certification operators to understand the costs involved in developing EPDs and PCRs. According to Tasaki et al. (2017), comparability and harmonization are related in terms of cost used in developing EPDs and PCRs, for example, the use of harmonization could bring about methods which are streamlined and bring about a decrement in the cost of product performance assessment and certification but a case whereby the need for high-level comparison of products arises will bring about an increment in the cost of product performance assessment and certification. Life-cycle assessment on sustainability and social behavior can bring about an increase in the number of impact categories needed for future assessment of EPDs thereby increasing the cost of product performance assessment and certification. Their result showed that the age of the EPD program or scheme affected the cost of the EPD development process i.e. the older the program, the lower the cost, this is as a result of the assumption that the operators of the program would lower their cost as they become experienced due to the support provided by their analysis. Their observation showed that EPD programs were higher, lower in costs with a lesser workforce in Asia as compared to Europe and the U.S. signifying that the more the available EPD programs existing for use, the lesser the cost, workload and workforce needed for the implementation of EPD. In summary, the basic EPD information on cost, comparability and coverage helps to improve processes involved in EPD and PCR.

3 Impacts of Energy Footprint Evaluation Methods

Evaluation methods of energy footprints are now revolutionizing discussions on climate change due to the great potentials they possess for the achievement of a reduction in energy emissions.

Energy footprint evaluation methods ensure the extension of mitigation options. Evaluation methods have decision-making options and alternatives which would ensure that emissions and footprints are reduced. This includes options of comparability, harmonization, taxation (in a situation where organizations are charged per emission rate), and adjustment on pattern and style of consumption which are widely known to reduce environmental levels of emission (Khan et al. 2009; Kiesecker et al. 2010).

Evaluation methods are crucial in increasing the level of participation in reduction of greenhouse gases. Mitigation options like taxation increase the level of priority which individuals and organizations attach to carbon emissions as they are taxed per rate of emission. This, therefore, ensures that people participate in the process of ensuring the reduction of greenhouse gases as money which they earn or work for becomes involved (Lan et al. 2016).

Results obtained from evaluation methods increase awareness on levels of emissions, it encourages governments and organizations to implement policies on competitiveness by providing benefits and luring packages which would increase participation in the climate change control process. The management features of evaluation method such as databases support business and government policy initiatives e.g. waste management strategies which are aimed at meeting the aims and objectives of emission level reduction. Policies and mechanisms are usually set up as a response to results obtained from evaluation methods to ensure the development of clean trade within a nation and between nations (Shan et al. 2014).

Environmental extended input-output analysis allows a quick assessment of drivers of upstream environment impacts and embodied impact of bilateral or multilateral trade, this allows for the development of new patterns of analysis from the perspective of consumption activities and not production activities to evaluate causes of global environmental impact of degradation and use of resources. This allows for tracking of resource and pollutant flow across the globalized economy and provides options for the reduction of impacts from the demands of human consumption which grows consistently (Wiedmann 2009).

4 Developing an Energy Footprint Evaluation Model

A model for evaluation of energy footprints is a representation of procedures, processes and conditions aimed at ensuring a proper evaluation of energy footprints. Models of evaluation are divided into two, namely: substance models and economic models.

(i) Substance Models of evaluation: This model involves processes such as achieving goals, side-effects of applying the model, focus on clients, focus on stakeholders, and peer review of models. Substance models are majorly oriented in achieving results and they address issues of interventions on outcome and output.
(ii) Economic Models of Evaluation: This model involves processes such as productivity, costs and benefits, and cost-effectiveness. Economic models are majorly oriented in achieving results and they address issues of interventions on cost by integrating all aspects of public interventions on substance and cost.

Burger and Gochfeld (2012) in their study, proposed a model for evaluating ecological footprints of energy sources. They made a proposal for the expansion of ecological footprints into sub-divided sections of the environment namely: underground (sub-surface or ground water), surface (land), airshed and atmosphere as different energy areas use the environment space differently. They also proposed a three-dimensional space for sub-surfaces (underground) to easily account for water rocks which energy sources have impact on. Burger and Gochfeld (2012) suggested the inclusion of extra considerations for the development of onshore and offshore activities in the sub-divided environmental sections, this is necessary as the consequences and effects of the development of energy resources on each sub-divided environmental section of the ecology. Research via collaboration with the public to get opinions on activities related to the sub-divided environmental sections are very crucial for the design of monitoring schemes used for the evaluation of energy footprints. Monitoring and assessing all types of renewable and non-renewable energy facilities is very necessary as it enables government and all members of the public compare and contrast all sources of energy for the development of energy policies that are sustainable. Burger and Gochfeld (2012), therefore, proposed a continuous and effective monitoring of both renewable and non-renewable energy sources (e.g. wind, solar, geothermal, hydro etc.) in all sub-divided sections of the environment including sediment, air, soil and water. The model is, therefore, a representation of how monitoring and division of the environment into four spaces allows a proper evaluation of energy footprints.

In a study by Radu et al. (2013) for the development of an energy footprint calculation model for assessing emissions of greenhouse gases that are obtained from large-scale projects. The model takes into account all types of emission (upstream and downstream, direct and indirect) for the calculation of carbon footprints. The model is very suitable for individuals with little knowledge on energy emissions and is not sufficiently aware of the importance of emissions from greenhouse gases. The proposed model took adequate consideration of both the period of implementation and period of operation. The analysis of the emissions included greenhouse gases such as SF_6, $CO_2, N_2O, PFC, NF_3, HFC$ and CH_4. The model can be applied in activities such as (i) Food processing and cultivation, (ii) Agricultural and hunting operations, (iii) Water collection, purification and distribution, (iv) Construction of buildings (v) Hotel and tourism operations, (vi) Restaurant operations (vii) Education (viii) Research and Development.

The model is centered around the efficiency principle (which aims at equaling benefits with cost) consisting of a baseline scenario (which represents an alternative to the project or situation which the model is applied to a situation whereby that project or scenario is absent or fails to occur) and the project scenario (which represents the application of the model in the project itself). The model also accounted for absolute emission (involved in the project scenario), baseline emissions (involved in the baseline scenario) and the relative emissions (the difference between the baseline scenario and project scenario) which gives positive or negative values where positive values represent high-levels of emissions and negative values represent low-levels of emissions. The model proposes the division of activities into low-level emission activities, medium-level emission activities and high-level emission activities for easy implementation of the calculation methodology. Data required to calculate the footprints of carbon in each operation and activity can be obtained from the life-cycle analysis of the project, the flow or use of technology, a detailed and comprehensive description of the processes of production and business plans. Provision of data is the responsibility of specialists and technologists in the field related to the project. The model could exist as a programmatic documentation for the creation of guidelines aimed at evaluating emissions of energy and ensuring development of sustainability in the environment.

In summary, developing a model for the evaluation of energy footprints involve the following

(i) Setting aims and objectives. This involves stating the reasons for creating the model e.g. for reducing emission levels, monitoring emission levels, funding purposes etc.
(ii) Building a flowchart showing how the model will be represented and organized.
(iii) Setting boundary and modelling conditions which serve as a basis for the functioning of the model. These conditions are set after careful research and investigation of activities related to the emission of greenhouse gases.
(iv) Collection of data for footprint calculation which would be used to test the model for efficiency and accuracy. The types of data to be obtained include activity data and emission factor data. Activity data is the data concerning greenhouse gas emissions obtained from different activities and operation while emission factor data is the data on various average rates of emission for a given greenhouse gas obtained from various sources which depend on different activities. Generally, carbon footprints can be calculated with four formulas on use of energy, transportation, food items, products and services.

(a) Carbon footprint ($kg\,CO_2/period$) = Activity data on energy use (based on quality and capacity) × $Emission\,factor(in\,kg\,CO_2)$.

The energy use could be in kWh for electricity use, therms for use of natural gas, litres (for use of liquefied petroleum gas, fuel oil and water), and kg for waste. The footprint is evaluated for a time period which could be in weeks, months or years.

(b) Carbon footprint ($kgCO_2$/period) = Activity data on distance of transportation (based on quality and capacity) × Emission factor (in $kgCO_2$).

Where distance of transportation (either road, rail, water or air) all in kilometers (km). The footprint is evaluated for a time period which could be in weeks, months or years.

(c) Carbon footprint (kg CO_2/period) = Activity data on food consumption (based on quality and capacity) × Emission factor (in kg CO_2).

Where food consumption is in kilocalories (kcal). Items like meat, dairy, cereals, vegetables, fruits, oils, snacks and drinks are all emission agents.

(d) Carbon footprint (kg $(greenhousegas)/$ period) = Activity data on products and services (based on quality and capacity) × Emission factor (in kg (greenhouse gas)).

Where the value of the products and services are represented in monetary terms of currency. Products and services such as clothing, housing, education, medicine, communication etc. are all greenhouse gas emission agents.

(v) Application of the model to a real-life scenario.
(vi) Examination of results for comparison and decision-making.

5 Tools for Energy Footprint Evaluation

Energy footprint tools have proven to be very useful due to their application of technology, they are useful aids in commercial, manufacturing and institutional facilities for tracking energy consumption, significant end-use of energy and factors connected to energy use thereby bringing precision to efforts of environmental sustainability. Tools used for the evaluation of energy footprint track environmental footprint, social footprint, economic footprint, combined footprints and composite footprints. Various software tools for calculating energy footprints exist and vary in terms of operation. They range from carbon footprint measurement tools, lifecycle assessment (LCA) tools and greenhouse gases (GHG) measurement tools. Energy footprints evaluation tools can be categorized into three namely: Calculators, Protocols and guidelines, and Process-based models.

5.1 Energy Footprint Calculators

Energy footprint calculators are tools used for the estimation of energy footprints by making use of assumptions and sources. They exist in many forms both in software and hardware. Some examples of popular carbon calculators include SimaPro, GaBi, BEES, CASBEE, WRATE, Athena etc.

West et al. (2016) in their study developed a household-level footprint calculator named REAP Petite which links local consumption to global greenhouse gas emissions. The calculator evaluates footprints consisting of emissions coming directly from fuel burning activities and emissions of full supply chain coming indirectly from the demand purchases of goods and services. The calculator uses survey-based data to properly adjust the estimates of separated neighborhood footprints which would ensure a unique calculation of individual impact of footprints by comparing the individual impact with a known national average. The REAP Petite calculator reports product footprints by using standard industrial classifications of various economic sectors to evaluate national level emissions and consumption-based government emissions. Geodemographic profiling data is used by the tool to determine profile of household expenditures which is based on Classification of Individual Consumption According to Purpose (COICOP) classification. REAP petite uses historic data to measure the reduction potential of household emissions and impact. The tool is very valuable in lowering carbon emissions through monitoring the effectiveness of community-based interventions. The REAP petite tool employs a bottom-up footprinting method to properly determine if predicted emission levels of household and communities are accurate thereby encouraging policies targeted at reducing emissions. REAP petite enables the comparison of footprints by allowing the users to compare the impacts of their energy use with that of other individuals and communities. West et al. (2016) stressed the need for footprint tools to be checked regularly and updated to make up for lapses which may occur in terms of accuracy and consistency.

Padgett et al. (2008) compared ten U.S based carbon emission calculators to understand their differences and similarities. Their analysis focused on three areas namely:

(i) Comparison of the factors considered to be the average behavior of Carbon emission causing components by different calculators;
(ii) Comparison of annual carbon emissions of households and entities evaluated by the calculators.
(iii) Existing opportunities for reducing the emissions of carbon emission entities.

It was observed that most calculators demand a small number of input characteristics to calculate carbon emissions in the household with consumption of electricity being a parameter used by all calculators that were examined. Most of the calculators which were examined provided background information pertaining to the relationship of carbon emissions and natural gas, carbon emissions and transportation, thereby providing conversion factors based on national averages in the

United States (U.S) for various end-use conversions. It was observed that the results of the calculators in terms of carbon emissions due to differences in factors of conversion, methods of calculation, estimates of behavior and other sources. It was also observed that many of the calculators failed to provide adequate information on their methods and estimates thereby making it difficult to check the relevance and accuracy of the calculations. Various methods used by electric power companies in the production of electricity lead to a variation in the results of different calculators when evaluating carbon emissions obtained from the consumption of electricity (the variation in the results of evaluating carbon emissions from consumption of electricity is as a result of the failure to provide adequate information on their methodologies and estimates). The lack of adequate information, therefore, suggests the need for greater transparency in the design of carbon emission calculators. Transparency in the design of calculators would ensure that users properly understand the calculator's methods of calculation and the obtained results thereby allowing the user choose a carbon emission calculator which suits the needs of the user. A calculator which is transparently designed will influence relevant bodies and agencies to increase their support in ensuring consistency and accuracy in carbon calculators through pressure coming from the marketing, social and regulatory environment. Carbon emission calculators ensure increasing awareness on carbon emissions and methods to reduce their influence in the public domain, this, therefore, calls for the need to ensure clarity, consistency and transparency in their design to achieve the numerous potential benefits they offer to the public.

Aichholzer et al. (2012) reviewed how carbon calculators (through CO_2 monitoring and feedback) and the participation of society combine to achieve local climate targets. Carbon calculators were described to act as an effective eco-feedback system as they can collect carbon data on individual/group behaviors and activities. Carbon calculators as eco-feedback provide lots of benefits as a learning tool which creates an enabling environment for users to review their energy consumption behaviors, system/equipment operation and the impacts on the environment through experimentation and continuous feedback. Carbon calculators can come in form of smart meters thereby saving cost and reducing negative environmental impacts due to quick and accurate operation characteristics of smart meters (Pratt et al. 2010). Carbon calculators can also function as online tools where individual/group energy consumption data, and other given functionalities are gathered, processed, stored and managed (Aichholzer et al. 2012). A Carbon calculator as an online (Internet) tool allows users to enter, process, calculate and continuously control their individual/group balances in CO_2. Online Carbon calculator tools provide extra functions like newsletters, accessible information on societal participation in reduction of carbon emissions in different localities and nations, hints/measures for energy saving, hints/measures for carbon emission reduction and offsetting, web forum where users can discuss personal experiences etc. to ensure effective operation of the online tool. To ensure that the online carbon calculator is simplified and transparent, the calculator should be concerned with four activities (namely: energy supply, nutrition, mobility and energy consumption) and calculation examples with references and sources should be provided to properly guide and direct the users. Aichholzer et al. (2012) observed that a good number of individuals and organizations (i) support the use of carbon calculators as an eco-feedback platform to bring about positive impact in climate conditions; (ii) support the calls for improvement for eco-feedback systems in terms of learning effects, creation of awareness and user guidance; (iii) agree that lots of benefits are involved in the collective participation of society in the eco-feedback platform; (iv) agree that the eco-feedback platform can cause positive changes in CO_2 balances thereby ensuring that local climate targets are achieved. Aichholzer et al. (2012) encouraged the participation of society in climate-related activities through digital means (carbon calculators) as this would result in large-scale information sharing, creation of awareness and collaboration in the creation and implementation of policy thereby ensuring sustainable activities for the climate and environment.

5.2 *Protocols and Guidelines*

Protocols and guidelines are agreed on frameworks which aid in the evaluation of energy footprints. They describe how energy footprints should be evaluated, where energy footprints should be evaluated and who is considered competent to evaluate energy footprints (Sovacool and Brown 2010). Some existing energy footprint protocols and guidelines include

(i) Kyoto protocol which are sets of agreements passed in the year 1997 in response to the threats posed by climate change to reduce emissions by an average level of 5.2%,
(ii) Global reporting initiative which sets out reporting requirements for emissions and footprints,
(iii) Carbon disclosure project which assists and guides corporations and cities in accounting and disclosing emissions,
(iv) Greenhouse gas protocol which provides standards and guidance on tracking, accounting and reporting emissions.

Aichele and Felbermayr (2012) evaluated the effect of Kyoto protocol on carbon footprints and emissions using an estimation strategy called differences-in-differences IV. Using every aspect of the Kyoto protocol as a guideline, data on high-quality input-output tables and coefficients on sectoral emissions from official sources of countries signed on to the Kyoto protocol was used to calculate carbon footprints. They used the carbon content of trade (which is the total amount of CO_2 *emissions* obtained from every national environmental activity). Aichele and Felbermayr (2012) defined a country's (c) carbon footprint at time t ($FP_{c,t}$) as

$$FP_{c,t} = A_{i,t} + AIA_{i,t} \tag{4}$$

where $A_{i,t}$ is the amount of CO_2 emissions obtained from domestic activities in the country.

$AIA_{i,t}$ is the amount of CO_2 emissions obtained from importing activities in the country which can be accounted for by tracking the global chain of production for all imported products and components from the respective countries which they originated from.

The differences-in-differences strategy involves estimating the effect of carbon emission in countries signed on to the Kyoto protocol. Carbon emission data is measured from both treatment and control periods (which could either be chosen from variables geographic location, endowments with different types of energy resources, preferences of the representative consumers, climatic conditions etc.). The treatment period is the period of time in which the protocol has been ratified or has been in existence in a particular country while the control period involves the post-period and pre-period which involves the period of time before the protocol was ratified and the period of time after the protocol ceases to exist in the country. The difference in measured carbon emission between treatment and control period is used to infer the impact of the Kyoto protocol on carbon emission in the country (Xu and Lin 2016; Zhang et al. 2011).

Hillman and Ramaswami (2010) proposed a guideline named "hybrid life-cycle-based trans-boundary greenhouse gas emission footprints" for evaluating greenhouse gas emission in large cities. This involves adding specified and agreed limits of end-use energy within cities (inbound) and energy demand for carrying out day to day activities within and across cities (cross-bound) e.g. transportation, housing (cement), fuels, water etc. to determine emissions from applicable and relevant greenhouse gases. Using the hybrid demand-focused approach, the inbound energy use (emissions) and cross-bound energy demands (contributions) are computed, this involves adding together the flows of material and energy which are multiplied by specific greenhouse gas factors of emission which is summarized in Eq. 2 below

$$\begin{aligned}\mathrm{GHG} = {} & \{(\sum Use\ of\ Energy \times FE) + (\sum M_{waste} \times FE)\} \\ & + \{(\sum UMP \times FE_{LCA}) + (\sum EE_{fuel} \times FE_{W-W})\}\end{aligned} \tag{5}$$

Inbound use of energy involves the use of natural gas, petrol and electricity for activities involving industries, commercial hubs and residences within a city, FE represents the factor of emission obtained from international protocols and agreements, UMP represents urban material production across cities (e.g. gasoline production, and raw materials for producing food, cement and housing) while FE_{LCA} uses a factor of emission evaluated from databases of life-cycle analysis (LCA). EE_{fuel} Represents energy emissions from using fuel e.g. driving gas-powered vehicles, cooking with gas cookers etc. while FE_{W-W} represents the factor of emission obtained from the well—to—wheel analysis which involves full fuel cycle from production to end-use. The annual residential, commercial and industrial end-use data can be obtained from the billing data of local utilities which are considered to be a data of high-quality. To allow proper, accurate and consistent comparison of GHG emission across cities, the emissions computed for a particular city should be normalized on a per capita scale and should be compared to the known national per capita emission of the country where the cities are located. It is, therefore, shown that the combination of a proper and step-by-step accounting of all inflows and outflows of material energy to a city and life-cycle analysis (LCA) makes a very accurate tool for computing GHG emission footprints.

Robinson et al. (2018) outlined key elements which are useful for the standardization of routine processes of carbon footprinting. They reviewed, reconciled and evaluated the footprinting steps involved in reputable guidelines and protocols to propose a reformed guideline which is universally acceptable and standardized for evaluating footprints in higher institutions. The design of the proposed universal guideline is centered on research questions which include

(i) Which are the best emission sources in higher education institutions?
(ii) Are there known difficulties in calculating carbon footprint data and if there are, what needs to be improved?
(iii) Would introducing sector targets which push institutions to calculate and reduce indirect emissions (which includes both upstream and downstream emissions) be a wise idea?
(iv) Would the proposed universal standard tool have positive effects on the environmental sector by ensuring it edges closer to reaching carbon management goals?

Getting answers to these questions is vital in determining the universality and acceptability of the proposed guideline as it assists the designers by getting useful human information which would ensure a universally acceptable guideline. The routine carbon footprinting procedure which involves

(i) Identifying emissions from facilities which ought to be accounted for,
(ii) Establishing the activities responsible for these emissions,
(iii) Collecting the necessary data, and finally,
(iv) Evaluating and verifying the obtained results, can be reformed into three steps which involves identifying emissions to be accounted for and the activities responsible for these emissions, (this is otherwise known as scoping), collection of operational and non-operational activity data, and applying the carbon equation to evaluate the carbon emissions (this is otherwise known as conceptualizing), reporting verified and detailed information to relevant stakeholders (this is otherwise known as communicating).

The reformed steps are useful in achieving positive results in carbon emission management as it involves stakeholders in the evaluation and future process of decision-making. All challenges involved in the reformation can be solved by using a criterion known as cut-off which establishes rules aimed at assisting environment practitioners in determining the activities which should be included and excluded from their footprint evaluation process thereby ensuring the credibility and accuracy of the obtained result. Some of the rules include

(i) Exclusion of indirect emissions (scope 3 emissions) from products and services which are paid (these products and services are external to the organization where the footprint evaluation is conducted i.e. products of other organizations),
(ii) Exclusion of activities which can be evaluated in other places,
(iii) Inclusion of activities which are significant and critical to geography and business i.e. direct emissions.

5.3 *Process-Based Models*

This involves a collection of processes and techniques aimed at evaluating footprints of environmental related activities. A process-based model should be (Bloch and Ranganthan 1991)

- Descriptive: Providing relevant information on several occurrences in a process
- Prescriptive: Providing established standards, guidelines and rules to achieve process performance
- Explanatory: Providing necessary explanations on all functions and challenges of the models.

Minx et al. (2009) gave an overview of a process-based input-output model for carbon footprinting which can be applied in many regions of the world. The model is based on areas such as emission drivers, supply chains, national emission inventories and trade, household consumption and lifestyles, emission sectors, sub-national emission inventories and organizations.

(i) On national inventories and trade, application of input-output models involves estimating the carbon footprints of a nation i.e. this involves analyzing a country's greenhouse gases balance of trade or assessment of leakage.
(ii) On carbon footprint drivers, the application of the input-output model involves changes in the composition and level of demand, changes in the size of households, efficiency of carbon in global production and changes in the structure of global production.
(iii) On carbon footprint of sectors, Input-output models are properly suited for sectoral studies on carbon footprints as that they provide a detailed and accurate understanding of the direct and indirect emissions of greenhouse gases which are linked to production and consumption activities, this, therefore, makes it very important in tracking the performance of activities in benchmarking processes. In areas where input-output data are not specific most importantly in hybrid models, process data can be used to support input-output approaches to ensure proper application of input-output based modelling of footprinting carbon products, activities and related frameworks like life-cycle assessment of products and policy on integrated products.

(iv) On carbon footprint of supply chain, input-output model assess carbon footprints considering all aspects of production and supply.

(v) On carbon footprint of organizations, estimation of carbon footprints using the input-output models involves estimation of scope 1 emissions (emissions from sources within the organization), scope 2 emissions (emissions obtained indirectly e.g. through organization's purchase of electricity) and scope 3 emissions (emissions obtained indirectly from the organization's operations). To meet the challenge of accounting for scope 3 emissions, an organization can link its financial account with the input-output model in a hybrid network to provide a detailed, cost-effective and accurate technique of providing an estimate for scope 3 emissions.

(vi) On carbon footprint associated with household consumption and lifestyles, application of an input-output model involves steps and measures for tracking patterns of lifestyle which determine emission rates, tracking progress in behavioral changes and removing barriers to behavioral changes which have an effect on emission rates (emission values are, therefore, obtained through continuous tracking). Analysis using input-output models reflects lifestyle in the way people make use of resources. Use of resources results in carbon emissions at different rates. An input—output model which is universal in characteristics is very important in tracing/tracking the contributions of different lifestyles of households and individuals to emissions of greenhouse gases contributions, the model assists in the creation of well-informed and creative intra-societal policies of climate change.

(vii) On carbon footprints of regions and local communities (sub-national emission inventories), input-output models evaluate footprints of small areas by combining information on activities of global production with information on activities of local consumption.

Input-output models like other types of process models their advantages and disadvantages, its usefulness in various applications varies with respect to some certain factors and key determinants like (Cuddington et al. 2013; Chin et al. 2004).

- Time of use: Depending on the scale of economic and environmental flow of emissions within a year, input-output tables determine the suitability of input-output models in different applications.
- Data type: Data used in input-output models are usually between averages of twelve months to few years old. When technology is required in processing of data, adequate and detailed information is needed to make the input-output models sufficient for carbon footprinting applications.
- Efforts of cost and work: Input-output models allows quick analysis and assessment of emissions with little amount of effort and low costs once the model is in good shape. This, therefore, means that an input-output model needs to be structured properly and effectively for it to be useful in any application.
- Detail and thoroughness: Analysis of available data determines the sufficiency of detail provided by input-output tables. Specific information on various

processes is important for determining the sufficiency of detail provided by input-output tables.

Baki and Makropoulos (2014) examined the application of a process modelling tool named Urban Water Optioneering Tool (UWOT) which evaluates the footprints of energy by creating a model of the interactions between water and energy in a water supply infrastructure. UWOT models all stages of a water cycle in urban communities (which includes source, water treatment, water distribution, use, wastewater collection and wastewater treatment). UWOT employs a technique whereby interventions of both demand-side management and supply side management in urban water system can be easily assessed through aggregates generated from signals of demand for water from household appliances. The tool is also capable of making simulations for different types of households and different types of water technologies. UWOT calculates total energy consumption by multiplying specific energy production and use by the volume of water (generated throughout the period of simulation). The calculation occurs in different modelled stages of the urban water cycle. The results obtained from the simulations are compared with historical data from water abstractions to make deductions and inferences in terms of measures and interventions which would impact in issues related to climate change. The tool also provides a means for calculating power used for water production by multiplying the power produced by the hydro-power plant (transformer, turbine and generator) which consists of the water density, acceleration due to gravity, flow in turbine, pressure from flow in turbine and total efficiency of the hydro-power plant. The calculation of power in the hydro-power plant provides accuracy in terms of model estimates of energy production through the capture of discharge effects and variations in water level. UWOT as a process modelling tool for evaluation of energy footprints helps identify the most effective interventions for water systems and infrastructures in urban communities. The tool encourages the development of measures for ensuring sustainable use of water by minimizing negative trade-offs between water and energy and also maximizing the combined benefits obtained from the interactions between water and energy.

Rosso et al. (2012) analyzed the impact of equilibrium and dynamic process modelling as a tool for energy footprint evaluation in wastewater treatment. They analyzed equilibrium and dynamic footprint modelling approaches to understand the duties of process components in the evaluation of energy footprints. The equilibrium analysis otherwise known as cumulative analysis involves provision of accounts on all activities related to power consumption. The cumulative approach also requires that engineering personnel such as plant operators, designers, modelers and managers have a professional level of knowledge on information about specific wastewater treatment sites. In the cumulative approach, a census on site equipment is necessary to get the contribution of each equipment to the site energy footprint. The total energy footprint using the cumulative approach can be summarized in Eq. 6

$$\text{TFPE} = \sum_{a=1}^{n} TFPE_a = \sum_{a=1}^{n} \sum_{b=1}^{m} n_b . p_b . \eta_b . t_b \qquad (6)$$

where

- TFPE represents footprint for energy in one operation a (in kWh)
- n_b represents number of equipment b
- p_b represents power use in a single equipment b (in kW)
- η_b represents efficiency of one equipment b
- t_b total operation time.

To calculate the power use and efficiency, field data and assumptions for process parameters are required. The dynamic modelling approach also has its process built around Eq. 6 but differs from the cumulative approach as $n_b, p_b \, and \, \eta_b$ depend on other process parameters and are time-varying in the dynamic modelling approach. The dynamic modelling approach requires detailed and proper measurement procedures for effective output in results. Footprint modelling is very beneficial as it provides energy footprint reduction opportunities, improves process operations and ensures savings in energy.

6 Developing an Energy Footprint Evaluation Tool

Developing an evaluation tool involves ensuring effective use of the knowledge, skills and techniques required for design as well as the outcomes expected in performance. Quality assurance processes are also important in the development of assessment tools as they provide checks and balances on components and frameworks making up the evaluation tool (Stöglehner 2003).

Huang et al. (2009) described the development of a tool for life-cycle assessment in maintenance and construction of asphalt pavements. The development process involved analysis of data, sensitivity of data, possibilities of recycling and assessment of the challenges encountered in the application of life-cycle assessment in the practice of pavement construction. The stages of development of the tool include

i. Defining the aim and scope of the life-cycle assessment study
ii. Developing a life-cycle inventory that contains a compiled list of important environmental challenges throughout the lifetime of the product
iii. Development of a platform for the calculation of life-cycle impact assessment
iv. Presentation of results for further analysis and comparison.

A life-cycle assessment tool is considered to be in proper working condition, if

- it's database and methodology are recognized internationally
- the data used for operation relevant and recent
- it has many variables that are useful in application
- It must be available for update and revision of formulas and features.

The life-cycle assessment tool developed in Huang et al. (2009) uses Microsoft excel spreadsheet which is made up of 5 worksheets for graphic presentation and calculation of unit inventory (which contains formulas for calculation and list of unit processes), construction pavement parameters, results of characterization (which contains processes and factors for characterization), process parameters and inventories of project. The life-cycle assessment tool finally goes through application in case studies and project to build up its scope and capacity. The application in different case studies will bring out findings, possible issues and challenges which when worked upon will contribute to the tool's advancement. Improvement on the encountered challenges should focus on communication skills with the user; sophistication level in terms of swiftness; effectiveness and efficiency; addition of data sources that are up to date and relevant; and proper result presentation in terms of data and result quality.

Lu et al. (2015) described the framework for the development of a carbon footprints evaluation tool for buildings using the building information modelling (BIM) technology for data visualization. The advantage this tool has over other building information modelling software tools is its potential to allow designers and architects quickly assess carbon footprints of buildings for decision-making, this, therefore, makes decision-making an important aspect of a carbon tool for footprinting. Two challenges which other building information software tools experience are

(i) The difficulty experienced in applying facility management to the life-cycle of buildings which makes it difficult to use visualization methods in revealing footprints of carbon.
(ii) The absence of a suitable form of standardization in architecture which makes evaluation of building carbon footprints in architecture difficult. The development of the tool makes provisions for solving these challenges by assisting the building information modelling software to select local carbon footprint data using evaluation theory of carbon footprint. Focus on the method of evaluation for carbon footprint and database for carbon footprints were identified as the key to achieving breakthrough in solving the challenges. Certain pre-conditions required for the development of the tool include.

- Identification of spots for carbon reduction by the designer (the designer is required to have an existing knowledge on carbon footprint consumption status throughout the process of design) and immediate modification of the design after spot identification
- Selection of parameters/variables for carbon footprint calculation occurring in different phases e.g. daily use of energy is used in the design phase for calculation of building energy footprints from lighting
- Planning of the regional characters which include review of regulation and method of classification, database and theory of evaluation.

The stages of development for the tool includes

i. Creation of applications and components to guide designers in choosing appropriate schemes and materials for making effective decisions and adjustments in design
ii. Dividing the evaluation system into three portions namely
    - Database for carbon footprints which gives records and values of emissions in transportation and production processes, it also provides the CO_2 equivalent of recycled and raw materials.
    - Building information modelling tools which include the building information modelling template which links building information modelling to assessment of building carbon footprints by enabling the building information modelling software to capture corresponding values directly and the subset data which is extracted from the database.
    - User interface which includes the Application Programming Interface (API) and components for guiding designers in decision-making.
iii. Implementing the project operation process which consists of

- Preliminary setup
- Model build
- Disclosure of carbon
- Adjustment of design.

The tool assists the designer in decision-making by sorting parameters of design in order of priority (this is known as automatic priority sorting), enabling comparison between chosen designs and alternative designs (this is known as multiple schemes parallel comparison), visualization of information on performance and real-time feedback for solutions on design.

Robert et al. (2002) gave some key features and procedures to ensure the development of an energy footprint evaluation tool. They noted two factors needed for consideration in the development of the tool which includes (i) the evaluation of the compliance of actions with the overall aims and objectives; (ii) monitoring the impact of the tool.

According to Robert et al. (2002), carrying out qualitative analysis involving environmental impacts on the results provided by the tool and dynamic analysis which involves effects of material flow on the system operation of the tool, life-cycle assessments, and ecological management analysis are important for the development of a footprint evaluation tool.

7 Impact of Energy Footprint Evaluation Tools

Evaluation tools are known to have long-lasting impacts both in the society in terms of lifestyles, environment (in terms of the level of emissions) and in decision-making. In many instances, comprehensive studies are very crucial for the continuous orientation and enlightening of the populace in the society. This helps in proper development of measures which encourages sustainability of the environment.

In reviewing a carbon calculator for educators, policymakers and managers, Zald et al. (2016) discussed some of the impacts carbon calculators have on the environment and society. The impacts of carbon calculators vary widely in terms of system representation, requirements of input data and outputs. Complex carbon calculators in the course of evaluating footprints of carbon, simulate the impact of disturbances posed by natural disasters such as wildfires, insect outbreaks, earthquakes etc. thereby triggering reductions in carbon storage. These tools offer measures on changes which need to be made to management processes which would have impacts on carbon emissions in landscapes which are prone to wildfires by reducing the influence of disaster-causing agents. Carbon calculators show how climate change influences the operation of carbon directly by changing the frequency and extent of occurrence of disaster-causing agents and changing carbon processes such as decomposition, soil respiration, growth and mortality.

Carbon calculators account for how management activities interact with effects of climate change by assessing present conditions of the environment which have crucial impacts on the dynamics of carbon. Carbon calculators also give users the opportunity of quickly estimating the dynamics of carbon with respect to management processes and activities in a reasonable fashion. The presence of information and tutorials in some carbon calculators have an impact on the users by improving their literacy in every aspect of carbon emission.

According to Peter et al. (2017), some carbon tools are also characterized by their inability to detect options of mitigation along the chain of production due to the high rates of uncertainty in predicting greenhouse gas emissions. This flaw in design negatively affects climate change initiatives and the desire to achieve an environment which is sustainable. Carbon emission evaluation tools impact positively on lands and the environment by fulfilling the educational purpose of creating awareness by providing farmers with information on climate change in relation to crop cultivation. Reporting tools for assessment of greenhouse gas emissions on landscape assist policymakers in making proposals on options for mitigation of greenhouse gases by providing features which allow comparison of results on the emission of greenhouse gases from different countries, this is as a result of effective and efficient database features provided by many carbon footprint assessment tools. Through evaluation of projects via the assessment of emissions of greenhouse gases obtained from different projects, carbon emission evaluation tools assist consultants, policymakers, technicians and non-governmental organizations in project management (e.g. systems of production and plans of reproduction) and innovations.

According to Ariyaratne and Moncaster (2014), many in-house tools designed and developed by many commercial and non-commercial institutions designed to calculate carbon emissions from buildings and environment in general lack the ability to adapt to multiple different conditions and are, therefore, inflexible which affects it's efficiency and effectiveness in performance, this was due to the absence of software in the tools which are capable of ensuring flexibility in many conditions and circumstances. A building information modelling (BIM) software is potentially able to incorporate in the design process, features which allow assessment of carbon footprint data in multiple conditions. Building information modelling (BIM) provides an enabling environment for the assessment of building designs in terms of sustainability. The possibility of adding plug-in features (which contain embodied data on carbon emissions) to the building information management (BIM) software is still being explored. Plug-in features could come in form of technical information on specific products from suppliers, addition of equations for automation of calculations on embodied carbon emission processes.

Fu et al. (2014) in their study, assessed the impact of applying life-cycle assessment tools in building construction. Life-cycle assessment tools are very important in buildings as they provide builders with the opportunity of analyzing materials which have been specified for carrying out building projects in the area of construction and also preparing alternative options in terms of required materials which could reduce preexisting levels of carbon emission in the project whilst having a positive impact on the cost of carrying out the project. Furthermore, they bring about an improvement in the connection between the project personnel and stakeholders (designer, contractor responsible for construction, laborers and owner of the project) in the very important areas of building and environmental sustainability. This would ensure a spread of knowledge and invaluable expertise on emissions of carbon from the designer and contractor down to the client thereby ensuring that the client is sufficiently aware of different alternatives and choices which they previously had no knowledge on. This is very crucial and vital to the goal of ensuring sustainable development as the client would be given the responsibility of taking part in the approach to achieve sustainable development in the society. Many Life-cycle assessment tools give room for assessing the impact of emissions both on the economy and environment with features such as economic performance score and environmental performance score, they also allow designers to select environmental friendly building products and provide analysis on cost-effectiveness. Carbon footprint evaluation tools also provide opportunities for reducing embodied emissions in buildings and the environment through the identification of materials which are either currently causing carbon emissions or potentially able to cause carbon emissions, many of the tools also have features which can identify materials that can be recycled as recycled materials are known to reduce emissions of carbon in buildings. In summary, life-cycle assessment tools provide support in terms of decision-making for ensuring sustainability in construction/building process and design of buildings.

8 Conclusion

The effects of allowing maintained levels, stable or increasing emissions in greenhouse gases have become overwhelmingly visible both in households, organizations and the world at large globally. This, therefore, shows the urgency for all nations around the world to make proper plans to ensure a future which is more sustainable in energy. Under these situations and conditions, any measure, technique or action taken to ensure the reduction in emission of greenhouse gases is considered commendable and should be given approval by governments and authorities through the imposition of such measures and actions as legislations and laws in the society. Many individuals and organization have made conclusions for societies and nations to make the achievement of sustainability in the environment a top priority to ensure a reduction in negative impacts of climate change on nature and future generations.

In this chapter, the methods and tools for the evaluation of energy footprints have been discussed extensively. It is very important for all members of society to have knowledge regarding tools and methods of evaluating footprints to support the agenda of ensuring sustainability in the environment. During the calculation of footprints, it is very important to take note of all types of emissions (types of footprint include direct and indirect, downstream and upstream) through the use of highly developed knowledge and sufficient resources for the provision of a proper overview on the impact of emissions on the environment. It is, therefore, very important for organizations and institutions to ensure that methods, tools and models designed for the evaluation of energy footprints are easy to understand and have features for comparison of data.

It is very crucial and important that processes such as monitoring of emissions, maintenance and development of sequestration capacity (through the regulation and promotion of areas that are considered safe and protected from emissions) are combined to strengthen the implementation of the sustainable development initiative. By ensuring that efficient systems are employed in planning for land use and building constructions are properly controlled and monitored, all cultural and natural resources will be effectively protected and will bring about huge reductions in emissions of greenhouse gases which will support and promote development which is sustainable.

In addition to the knowledge present in this chapter, it is important for more researches to be undertaken on how existing tools and methods for evaluation of energy footprints can be developed and applied to conservation of the natural environment through the control of human activities for the reduction of negative impacts and disasters.

References

Aichele R, Felbermayr G (2012) Kyoto and the carbon footprint of nations. J Environ Econ Manage 63(3):336–354

Aichholzer G, Allhutter D, Strauß S (2012) Using online carbon calculators for participation in local climate initiatives. Lecture notes in computer science (including subseries lecture notes in artificial intelligence and lecture notes in bioinformatics), vol 7444 LNCS, pp 85–96

American Carbon Registry (2011) Tool for estimation of stocks in carbon pools and emissions from emission sources. *Carbon N. Y.*, no. April, Version 10, pp 1–67

Ariyaratne CI, Moncaster AM (2014) Stand-alone calculation tools are not the answer to embodied carbon assessment. Energy Procedia 62:150–159

Baki S, Makropoulos C (2014) Tools for energy footprint assessment in urban water systems. Procedia Eng 89:548–556

Bicknell KB, Ball RJ, Cullen R, Bigsby HR (1998) New methodology for the ecological footprint with an application to the New Zealand economy. Ecol Econ 27(2):149–160

Bloch C, Ranganthan R (1991) Process based cost modeling. In: 1991 proceedings eleventh IEEE/CHMT international electronic manufacture technology symposium, pp 406–412

Bogeskär M, Carter A, Nevén C-O, Nuij R, Schmincke E, Stranddorf HK (2002) Evaluation of environmental product declaration schemes

Boitier B (2012) CO_2 emissions production-based accounting vs consumption : Insights from the WIOD databases. In: Final WIOD Conference Causes Consequences Glob., no. April, pp 1–23

Burger J, Gochfeld M (2012) A conceptual framework evaluating ecological footprints and monitoring renewable energy: wind, solar, hydro, and geothermal. Energy Power Eng 04(July): 303–314

Chin S, Kim K, Kim Y-S (2004) A process-based quality management information system. Autom Constr 13(2):241–259

Cuddington K, Fortin MJ, Gerber LR, Hastings A, Liebhold A, O'connor M, Ray C (2013) Process-based models are required to manage ecological systems in a changing world. Ecosphere 4(2)

Del Borghi A (2013) LCA and communication: environmental product declaration. Int J Life Cycle Assess 18(2):293–295

EC I, OECD (2008) System of national accounts

European Commission, IMF, OECD, and United Nations (1993) System of national accounts 1993. J Gov Inf, p 814

Fet AM, Skaar C, Michelsen O (2009) Product category rules and environmental product declarations as tools to promote sustainable products: experiences from a case study of furniture production. Clean Technol Environ Policy 11(2):201–207

Fu F, Luo H, Zhong H, Hill A (2014) Development of a carbon emission calculations system for optimizing building plan based on the LCA framework. Math Probl Eng 2014

Hillman T, Ramaswami A (2010) Greenhouse gas emission footprints and energy use benchmarks for eight U.S. cities. Environ Sci Technol 44(6):1902–1910

Huang Y, Bird R, Heidrich O (2009) Development of a life cycle assessment tool for construction and maintenance of asphalt pavements. J Clean Prod 17(2):283–296

Ingwersen WW, Stevenson MJ (2012) Can we compare the environmental performance of this product to that one? An update on the development of product category rules and future challenges toward alignment. J Clean Prod 24:102–108

Jones C, Kammen DM (2014) Spatial distribution of U.S. household carbon footprints reveals suburbanization undermines greenhouse gas benefits of urban population density. Environ Sci Technol 48(2):895–902

Joshi S (1999) Product environmental life-cycle assessment using input-output techniques. J Ind Ecol 3(2–3):95–120

Khan S, Khan MA, Hanjra MA, Mu J (2009) Pathways to reduce the environmental footprints of water and energy inputs in food production. Food Policy 34(2):141–149

Khasreen MM, Banfill PFG, Menzies GF (2009) Life-cycle assessment and the environmental impact of buildings: a review. Sustainability 1(3):674–701
Kiesecker JM, Copeland H, Pocewicz A, McKenney B (2010) Development by design: blending landscape level planning with the mitigation hierarchy. Front Ecol Environ 8(5):261–266
Kitzes J (2013) An introduction to environmentally-extended input-output analysis. Resources 2(4):489–503
Lan J, Malik A, Lenzen M, McBain D, Kanemoto K (2016) A structural decomposition analysis of global energy footprints. Appl Energy 163:436–451
Lenzen M, Murray SA (2001) A modified ecological footprint method and its application to Australia. Ecol Econ 37(2):229–255
Liu LC, Liang QM, Wang Q (2015) Accounting for China's regional carbon emissions in 2002 and 2007: production-based versus consumption-based principles. J Clean Prod 103:384–392
Lu CM, Chen JY, Pan CA, Jeng T (2015) A BIM tool for carbon footprint assessment of building design. In: CAADRIA 2015—20th international conference computer architecture design research Asia emerging express past, present future digital architecture, pp 447–456
Madrazo J, Clappier A, Belalcazar LC, Cuesta O, Contreras H, Golay F (2018) Screening differences between a local inventory and the emissions database for global atmospheric research (EDGAR). Sci Total Environ 631–632:934–941
Manzini R, Noci G, Ostinelli M, Pizzurno E (2006) Assessing environmental product declaration opportunities: a reference framework. Bus Strateg Environ 15(2):118–134
McGregor PG, Swales JK, Turner K (2008) The CO_2'trade balance' between Scotland and the rest of the UK: performing a multi-region environmental input-output analysis with limited data. Ecol Econ 66(4):662–673
Miller RE, Blair PD (2009) Input-output analysis: foundations and extensions, 2nd edn.
Minx JC, Wiedmann T, Wood R, Peters GP, Lenzen M, Owen A, Scott K, Barrett J, Hubacek K, Baiocchi G, Paul A, Dawkins E, Briggs J, Guan D, Suh S, Ackerman F (2009) Input–output analysis and carbon footprinting: an overview of applications. Econ Syst Res 21(3):187–216
Padgett JP, Steinemann AC, Clarke JH, Vandenbergh MP (2008) A comparison of carbon calculators. Environ Impact Assess Rev 28(2–3):106–115
Peter C, Helming K, Nendel C (2017) Do greenhouse gas emission calculations from energy crop cultivation reflect actual agricultural management practices? A review of carbon footprint calculators. Renew Sustain Energy Rev 67:461–476
Pratt RG, Balducci PJ, Gerkensmeyer C, Katipamula S, Kintner-Meyer MCW, Sanquist TF, Schneider KP, Secrest TJ (2010) The smart grid : an estimation of the energy and CO_2 benefits. Contract, no., pp 1–172
Radu AL, Scrieciu MA, Caracota DM (2013) Carbon footprint analysis: towards a projects evaluation model for promoting sustainable development. Procedia Econ Financ 6:353–363
Rebitzer G, Ekvall T, Frischknecht R, Hunkeler D, Norris G, Rydberg T, Schmidt WP, Suh S, Weidema BP, Pennington DW (2004) Life cycle assessment part 1: framework, goal and scope definition, inventory analysis, and applications. Environ Int 30(5):701–720
Robèrt KH, Schmidt-Bleek B, Aloisi De Larderel J, Basile G, Jansen JL, Kuehr R, Price Thomas P, Suzuki M, Hawken P, Wackernagel M (2002) Strategic sustainable development—selection, design and synergies of applied tools. J Clean Prod 10(3):197–214
Robinson OJ, Tewkesbury A, Kemp S, Williams ID (2018) Towards a universal carbon footprint standard: a case study of carbon management at universities. J Clean Prod 172:4435–4455
Rosso D, Jiang L, Sobhani R, Wett B (2012) Energy footprint modelling: a tool for process optimization in large wastewater treatment plants. Water Pract Technol 1(7):1–8
Schaltegger S, Csutora M (2012) Carbon accounting for sustainability and management. Status quo and challenges. J Clean Prod 36:1–16
Shan Y, Čuček L, Varbanov PS, Klemeš JJ, Pan K, Zhu H (2014) Footprints evaluation of China's coal supply chains. Comput Aided Chem Eng 33:1879–1884
Simonen K (2014) Life cycle assessment. Life cycle assess, pp 1–159
Sovacool BK, Brown MA (2010) Twelve metropolitan carbon footprints: a preliminary comparative global assessment. Energy Policy 38(9):4856–4869

Stadler K, Wood R, Bulavskaya T, Södersten C-J, De Koning A, Tukker A (2014) EXIOBASE 3—compilation and analysis of an EE MRIO time-series in current and constant prices. In: 24th International Input-Output Conference, 2014, p 514

Stöglehner G (2003) Ecological footprint - a tool for assessing sustainable energy supplies. J Clean Prod 11(3):267–277

Tasaki T, Shobatake K, Nakajima K, Dalhammar C (2017) International survey of the costs of assessment for environmental product declarations. Procedia CIRP 61:727–731

Ten Raa T (2006) The economics of input-output analysis

Tukker A, Poliakov E, Heijungs R, Hawkins T, Neuwahl F, Rueda-Cantuche JM, Giljum S, Moll S, Oosterhaven J, Bouwmeester M (2009) Towards a global multi-regional environmentally extended input-output database. Ecol Econ 68(7):1928–1937

Turner K, Lenzen M, Wiedmann T, Barrett J (2007) Examining the global environmental impact of regional consumption activities - part 1: a technical note on combining input-output and ecological footprint analysis. Ecol Econ 62(1):37–44

UNEP/SETAC Life cycle Initiative (2005) Life cycle approaches : the road from analysis to practice. Assessment, p 89

Walther G-R, Berger S, Sykes MT (2005) An ecological 'footprint' of climate change. Proc Biol Sci 272(1571):1427–1432

West SE, Owen A, Axelsson K, West CD (2016) Evaluating the use of a carbon footprint calculator: communicating impacts of consumption at household level and exploring mitigation options. J Ind Ecol 20(3):396–409

Wiedmann T (2009) A first empirical comparison of energy footprints embodied in trade—MRIO versus PLUM. Ecol Econ 68(7):1975–1990

Wright LA, Kemp S, Williams I (2011) 'Carbon footprinting': towards a universally accepted definition. Carbon Manag 2(1):61–72

Xu B, Lin B (2016) Regional differences of pollution emissions in China: contributing factors and mitigation strategies. J Clean Prod 112:1454–1463

Zackrisson M, Rocha C, Christiansen K, Jarnehammar A (2008) Stepwise environmental product declarations: ten SME case studies. J Clean Prod 16(17):1872–1886

Zald HSJ, Spies TA, Harmon ME, Twery MJ (2016) Forest carbon calculators: a review for managers, policymakers, and educators. J For 114(2):134–143

Zhang Y, Zhang J, Yang Z, Li S (2011) Regional differences in the factors that influence China's energy-related carbon emissions, and potential mitigation strategies. Energy Policy 39(12): 7712–7718

Establishment of Electrical Energy Benchmarking Protocol for the Assessment of the Carbon Emissions in Hotel Industry

O. M. Babatunde, P. O. Oluseyi, I. H. Denwigwe and T. J. Akin-Adeniyi

Abstract The hotel industry consumes quite a sizeable quantum of energy. The industry stakeholders contend with energy related challenges; this includes high cost of service provision as a fall out of electricity price as well as poor energy policy. More so, this industry remains an unattractive subsector to both the energy players and succeeding governments (in Nigeria). Due to the great importance of this industry to commerce, tourism and hospitality; it is thus necessary to conduct its energy performance analysis. This will indirectly provide a template for an efficient utilization of energy in the industry. Though there are no creditable national records on energy consumption in this industry in Nigeria, but a well-calibrated energy data treatment method is developed to establish an acceptable protocol using relevant statistical tools. These methods are tested on an empirically selected number of hotels which are actively involved in business activities in Lagos metropolis. The results obtained clearly established that a number of building parameters have a set of measurable influence on energy use, as well as carbon emission, in the hotels. To support this, it is discovered that the relationship between the total floor area (footprint) and number of employees has the highest influence on energy utilization in the hotels.

Keywords Energy benchmark · Hotel industry · Statistical tools
Emission · Energy use intensity

1 Introduction

Hotel facilities have been in the history for several years since the pre-medieval ages. It has been called various names at different times. This includes such names as inn, hotel, tavern, bar, hostelry, taproom, motel, etc. The importance of this

O. M. Babatunde · P. O. Oluseyi (✉) · I. H. Denwigwe · T. J. Akin-Adeniyi
Department of Electrical and Electronics Engineering, University of Lagos,
Lagos, Nigeria
e-mail: poluseyi@unilag.edu.ng

S. S. Muthu (ed.), *Energy Footprints of the Bio-refinery, Hotel, and Building Sectors*,
Environmental Footprints and Eco-design of Products and Processes,
https://doi.org/10.1007/978-981-13-2466-6_3

industry is quite enormous, just as it can serve as a great source of income as well as employment provision. It also supports the tourism facilities. Its importance was recognized in the early years as far back as 15 BC when it was used for accommodating those travellers in today's Middle East. In Africa, there is a rich cultural history of the existence and importance of inns in the information available on the commerce and development of Egypt and Ethiopia. In the same vein, the existence of the hotel industry in Nigeria dates back to early 1940s.

Though there is very limited information on the history of hotels in Nigeria. A version of the story has it that the first resemblance of hotels was started as the Nigerian Railway's Catering Services on 10th March 1944 which later metamorphosed to the Nigerian Hotels Limited in 1946. Since Lagos was the operations headquarters of the colonial masters of Nigeria, there was a number of catering services and bars that were raised to take care of the hospitality aspects of government activities. These were in form of Guest Houses. In Lagos, the group of Guest Houses of the Federal Government of Nigeria was later handed over to a private business mogul that later changed the group of Guest Houses to the Ikoyi Hotels in Lagos. Yet another version had it that the first private hotel in Nigeria was Olympic Hotel which started in 1944; followed later by the Grand Hotels both in Lagos. After which the hospitality industry began to spread to other parts of the federation. In other words, the appreciable activities and influence of hotel industry in Nigeria will only be exhaustively considered by carrying out the study of its activities on its operation in Lagos.

Meanwhile, the growth of the industry cannot be totally disconnected from the fact that gradual noticeable development in the economy and visible presence of the whites in the country facilitated earlier stated expansion. The industry, though crucial to the national life, has received very little consideration in terms of the investment of resources on its activities by the academic and research experts. For instance, only very few higher institutions in Nigeria have a curriculum for Hotels Management. This has impacted negatively on the research that could encourage in the development and dynamic promotion of the industry through implementation of research outcomes.

Due to the current and pressing need for energy policy in the industry; unlike the previous governments; there is a growing interest and sensitivity to encourage the adoption of the minimum benchmark level of comfort based on the classification of hotels on the star-graded nomenclature. Though, there has been dividing opinions that the star-based classification of hotels in Nigeria has not been well defined which the hospitality industry has maintained that it is as a results of the proliferation of small and medium scale hotels without proper government approval. So also, it has been recorded elsewhere that a number of these hotels strive to meet the set standard but for the funding and periodic maintenance of facilities; which has a heavy toll on the generated revenue. Thus, the industry has the notion that since a number of sectors of the economy has received financial assistance from the government thus the hospitality industry, with special interest in the hotels, should be considered for this kind of favour.

As stated earlier the hotel industry in Nigeria; like many other sectors, suffered from paucity of policy that can discourage quackery and place premium on professionalism and best practices in service deliveries. This has been attributed to a number of factors among which is the building functionalities and operations in the industry. For instance, the building that is meant for residential purpose which is later converted and retrofitted to accommodate a hotel may not be able to take care of the vital requirements that are meant to be made available in such facilities. This includes such essential trade mark of the industry as energy profiling, purpose-built lavatories and other engineering service provision in the building. All these impede on the efficient energy and exergy outputs of the facilities.

Since the hotel industry is a very crucial aspect of hospitality and social development of many countries of the world. It has been noted in several countries that the hotel industry greatly contributes to the growth of the global economy by providing services of sightseeing, accommodation, and all other services related to culture and tourism. Moreover, the hotel industry is not only known to provide employment but is also very important in the provision of economic returns for a nation in terms of foreign exchange (if properly harnessed). The hospitality industry also provides informal education to people of different races, ethnicities and cultures; by providing the platforms for interaction among various nationalities. This therefore put pay to the role which the hospitality industry plays in developing the economy of a nation.

For the hotel industry to achieve the above, the energy supply must be adequately and abundantly available. Due to the peculiarity of the energy crisis in Nigeria; there is a prevalence of energy poverty in the sense that the poor energy supply/energy demand gap. This inability of the utility to provide the energy demand has resulted in the captive provision of energy to meet both production of goods and services in Nigeria. So there is need for the entrepreneurs in the hotel industry, like elsewhere in the national economic value chain, to make adequate provision for alternative energy supply.

This is further buttressed by the fact that the hotel subsector consumes significant amount of energy and energy-related resources in its daily operations due to its peculiar architecture and building functionalities as a commercial sector of the energy classification (Xydis et al. 2009; Bohdanowicz and Martinac 2007). This is further acerbated by the fact that many other commercial buildings have specific operational hours; while it is quite uncommon as well as ambiguous to define the operating hours for this peculiar industry. This thus makes the industry an energy-intensive sub-sector of the economy. For instance; while restaurants may close as late as the hour of 12 midnight, facilities such as the guest rooms, lobby, lifts, and pool side are provided with day-long illumination with some of them also being enclosed spaces with the conditioning of the spaces throughout the entire day. Hence, the above-mentioned factors contribute to the observed complexities in estimating both the energy use pattern and indicators in hotel buildings.

As stated earlier; the hotel building consists of different functional spaces such as: the guest rooms, restaurants, laundry, kitchen, gym, swimming pools, car parks, business centers, and sometimes multipurpose halls as well as auditoria. As classified by a group of researchers, these space zones can be categorized as: guest rooms, public areas and service areas (Bohdanowicz et al. 2001). For instance, the guest rooms are of distinct volumes with varying electrical load demands depending on the type and size of the room. For obvious reasons, the public spaces consume more energy due to the number of installed air handling units. Example of such is the cooling of elevators, corridors and staircases. Furthermore, the service areas usually demand high energy due to the cooling requirement for the energy consuming loads in such areas. All the three categories of space zones consume quite a lot of energy, especially electricity so as to remain functional, habitable and comfortable for customers. While satisfying the customers in terms of comfort, there is also need to investigate the energy use pattern of the hotel building so as to determine the energy consumption level as well as propose and implement measures to improve energy efficiency of this category of buildings. The study of hotels' energy consumption classification footprint and performance is therefore imperative in encouraging energy efficiency without reduction in the effective hotel administrative and commercial operation.

The rest of this chapter is arranged as follows: Sect. 2 presents the overview of related studies, Sect. 3 provides information on the background study of the hotel industry in Nigeria, while Sect. 4 establishes the requirements for the building energy audit as well as building information framework and Sect. 5 showcases the methodology with the presentation of related results, with Sect. 6 advancing the concluding remarks with suggestion on the relevant future research horizons.

2 Literature Review

The Hotels' energy consumption depends on diverse factors which are classified into four broad categories, namely: technical, operational, metrological and managerial factors. The investigation into this topical issue has attracted attention of many researchers (mainly hotel buildings) over the years. For instance, Y. Xin et al. 2012. modeled energy scenario in hotel with empirical information on the energy consumption benchmark for four and five star luxury hotels in Hainan Province . In line with the report on the building characteristics data acquired, correlation analysis and degree day method, were applied to establish the energy consumption quota and normalized energy use index respectively. With an R^2 of 0.879, the, energy consumption per unit area was discovered to be the most correlated index to characterize the energy consumption level. In other words, depending on the hotel type, the Energy Utilization Index (EUI) of the Hainan hotels is estimated to span within the range of 69.23 and 96.75 kWh/m^2. Using a multiple variable regression analysis, it was adduced that hotel standard, total hotel floor area, number of guest-nights sold and number of food covers sold could be correlated with the energy and water use in these facilities without normalization with regard to the weather/climate data (Bohdanowicz and Martinac 2007). The study further discovered that a high heterogeneity of hotel characteristics necessitates the sub-classification of hotels into specific set in order to obtain reliable consumption models and benchmarks for the hotels.

In order to identify the energy conservation and carbon reduction (ECCR) indicators for the hotel industry in Taiwan, some authors conducted semi-structured interviews with cornucopia of the professionals and regulators in the hospitality sector such as selected senior hotel managers, environmental specialists/analysts as well as policy makers in the government (Teng et al. 2012). From the findings of this study; a total of 32 performance characteristics were identified and ranked in terms of the respective individual comparative significance to ECCR contributions. Furthermore, a variant of Analytic Hierarchical Process (AHP), Analytic Network Process (ANP) was adopted for treating the questionnaire so as to establish the relative weights of the benchmarks. The result of the analysis returned seven classifications in the ECCR framework, which are: communication and participation, top management commitment, energy, water, waste, building, and purchasing. Results from the study shows that the success of ECCR execution depends predominantly on management support as well as staff engagement. In another study, a regression-based benchmarking model which accounts for the difference in functional and operational features when hotels are compared based on their energy performance was established (Xuchao et al. 2010). Using standard procedure for corporate greenhouse gases (GHG) emission accounting, CO_2 emissions from the surveyed hotels were also estimated. Worker density and a dummy variable which is 1 for 3-star hotel and 0 for higher star rating were the most correlated variables that formed the basis of the EUI model.

In order to investigate the relative efficiency in energy utilization of hotels in Antalya region of Turkey, an evaluation of energy efficiency in 32 five-star hotels was carried using data envelopment analysis (DEA) (Önüt and Soner 2006). The results from this study indicate that eight hotels are efficient while the other twenty-four hotels are inefficient in terms of energy consumption/utilization. Hence, an ideal energy consumption protocol was proposed for the 24 inefficient hotels. R. Priyadarsini presented a study on energy performance of hotel buildings in Singapore (Priyadarsini et al. 2009). Using a data sample of 29 quality hotels, energy consumption data were analyzed to obtain the energy use indicators (EUI). The adopted Statistical approach shows a weak correlation analysis between electricity consumption and number of occupied rooms in the selected hotels- an indication that it is necessary to improve energy management at low occupancy rate. So also adoption of other fuzzy-related approach for energy efficiency measure in public building has been considered. Furthermore, Pearson correlations between hotel energy use intensity and proposed descriptive factors shows that three-star hotels differ from higher stars in energy use. The worker density and years after the last major energy retrofit were also found to be highly correlated to hotel building EUI. The annual average total energy use intensity (EUI) in these hotels is 427 kWh/m^2.

A study that explored Jordan's energy consumption in the tourist accommodation industry has also been presented (Ali et al. 2008). From this, it was discovered that hotel's classification is significant in elucidating variations, if long-term investments, is to reduce energy consumption by using energy efficient appliances. The Five- and four-star hotel categories were the most receptive hotels willing to adopt energy efficient appliances to reduce energy consumption while the managements of one-star hotels were reluctant to introduce energy saving appliances. Shiming and Burnet presented the relationship between the energy use intensity and the average annual occupancy rate for 16 Hong Kong hotels using regression analysis (Shiming and Burnett 2002). No specific relationship was established. The regression analysis results also indicated that electricity use in the hotels, under review, is correlated with both number of guests and outdoor air temperature, with the latter being the stronger indicator. Farrou et al. categorized hotels in Greece using the k-means algorithm controlled with the silhouette plot after normalization of the operational energy data (Farrou et al. 2012). Ninety hotel buildings were analyzed and well separated clusters were specified for the sample size. The investigation revealed that energy use varies significantly between and within clusters- an indication of its fuzzy rather than specific value quota for the type of building.

A report on the survey of energy and water use in 36 quality hotels in Hong Kong has been presented (Deng 2003). The Regression analysis was used on the surveyed data in order to correlate them to a number of hotel operational factors such as gross floor area, number of guestrooms and food cover. It was also used to search for suitable energy and water use performance explanatory indicators. The gross floor area is the most correlated indicator and was able to explain 65.5% of energy use in the hotels. A study of energy performance in 16 quality hotels in Hong Kong reported by Shiming and Burnet shows that average EUI based on unit floor area is not sufficient (Deng and Burnett 2000). Other explanatory indicators explored include: year of construction, hotel star rating and occupancy level. In order to adequately access the energy performance in the hotel buildings, the study suggested categorization of hotel building into two parts: guest floors and non-guest floors.

In a separate research conducted by two separate groups of researchers, no significant relationship was established between energy use and occupancy rate (Becken et al. 2001; Reddy et al. 1997). With a R^2 of 0.95, another study was able to acknowledge that an exponential relationship existed between monthly electricity consumption and the number of guests (Papamarcou and Kalogirou 2001). In a robust support of the above result, Yao et al. (2015) reported the difficulty in establishing a significant relationship between the EUI and occupancy rate. It further noted that the relationship between individual building energy use intensity does not strictly correlate with the hotel star rating. Consequently, from the above review, it is evident that in order to establish a meaningful relationship between EUIs and the performance indicators in hotel buildings, categorizing all hotels into different sub-class is necessary to reveal homogeneity. Table 1 shows the average EUI of hotels in different countries around the world.

A search of the literature revealed that only a study has so far been conducted on energy performance indicators in Nigerian Hotel buildings. In an earlier study, correlation analysis was adopted to examine the relationship between carbon footprint and normalized energy consumption (Oluseyi et al. 2016). Though significant correlation was established between the energy use and indicators such as: total floor area, number of guest rooms, equivalent number of guest rooms and

Table 1 Average EUI (kWh/m^2/year) for hotels worldwide

Location (year data was published)	Mean EUI (kWh/m^2/year)	Method	Source
Hong Kong (2000)	564	Not clear	Deng (2003)
Cyprus (2001)	272.6	Not specified	European and Energy (2001)
Greece (2001)	107.52	Not specified	European and Energy (2001)
Italy (2001)	364.4	Not specified	European and Energy (2001)
Portugal (2001)	269.5	Not specified	European and Energy (2001)
New Zealand (2001)	158.6	Not specified	European and Energy (2001)
Hong Kong (2001)	406	Not clear	Yik et al. (2001)
Hong Kong (2002)	564	Multiple variable regression analysis	Shiming and Burnett (2002)
Hong Kong (2002)	342.02	Regression analysis	Chan and Lam (2002)
Hong Kong (2003)	542	Regression analysis	Deng (2003)
Vietnam (2005)	115.75	Unclear	Kumar et al. (2005)
Turkey (2006)	388.8	Data envelopment analysis	Önüt and Soner (2006)
Europe (2007) {Hilton Hotels}	364.3	Multiple variable regression analysis	Bohdanowicz and Martinac (2007)
{Scandic Hotels}	285		
Singapore (2009, 2010)	427	Regression, Pearson correlation	Priyadarsini et al. (2009)
Balearic Islands {Spain} (2010)	147.45	Not specified	Rosselló-Batle et al. (2010)
UK (2011)	213		Filimonau et al. (2011)
Greece (2012)	182	k-means algorithm, silhouette plot	Farrou et al. (2012)
Taiwan (2012)	211.925	Regression	Wang (2012)
Hainan Province {China} (2013, 2012)	131.81, 82.99	Multiple variable regression analysis (normalized with degree-day), correlation analysis	Lu et al. (2013)
South Korea (2014)	346.22		Ryu et al. (2014)
Taiwan (2013)	269	Multiple-regression	Wang and Huang (2013)
Tunisia (2005)	271.45	Average/mean	Khemiri and Hassairi (2005)

(continued)

Table 1 (continued)

Location (year data was published)	Mean EUI (kWh/m^2/year)	Method	Source
Shanghai (2015)	279.8	Mean	Yao et al. (2015)
Nigeria (2016)	265.95	Correlation analysis	Oluseyi et al. (2016)

employee, there are possibility of deficiency because of non-classification of sampled hotels before analysis.

3 Hotel Business in Nigeria

The hotel industry in Nigeria dates back to the 1940s when the first commercial hotels (Grand hotel and Bristol hotel) came into existence. Prior to this period, travellers, either on vacation or work-related purposes, sought shelter in villages and hamlet after which they paid their hosts either little or no remuneration in return for the act of hospitality. This was unsafe and unreliable for both the hosts and the travellers seeking accommodation as there was no certainty that the travellers would be fortunate to meet someone who would be kind enough to provide hospitable accommodation. Furthermore, the hosts also had no reliable means of guaranteeing the safety of the unknown person. Gradually as development began to evolve, lodges were built by the colonial rulers to cater for their families and other foreign visitors/travellers who came from Britain and other foreign nations. Initially, most of these lodges were built close to the railway stations which were then the commonest means of transportation; for convenience and security. Immediately after the independence, the hotel sector in Nigeria grew sporadically and flourished due to technological, industrial and the socio-economic development experienced by the country. In such a way that the hotel patronage and occupancy rate grew rapidly in the mid-1970s due to the influx of expatiate to the country during the oil boom era (Abomeh 2012). In 1981, a survey carried out shows that the hospitality and tourism industry contributed 0.3% of the total export (Ibeh et al. 2016). According to this report, it was projected to rise to 4.5% by the year 2020. Going by this, it could be concluded that tourism activties and the need for hospitality were factors that led to the development of the hotel sector in Nigeria. To meet the taste of the various customers, star rated hotels are being put up majorly in commercially viable cities in Nigeria. This informed the reason for the concentration of hotels in Lagos state South-Western Nigeria. This is due to its strategic potentials, since Lagos was a former capital of the federal republic of Nigeria which hosts the busiest international airport in Nigeria, a major sea port of entry for goods, Nigerian and regional headquarters of multi-national companies and notable higher institutions of learning. This has made the state to be a major benefactor from the boom of tourism and hospitality industries. Majority of these global branded star rated hotels are located on the mainland near the international airport and higher institutions of learning, and on the island, close to the sea ports.

According to (Sanni 2009), the presence of few numbers of international hotels in Nigeria as compared to the contemporary elsewhere in the world is as a result of the perception of Nigeria being a high-risk location for business investments. This is more evidently exemplified, especially, in the hospitality sector where capital intensive and long-term business commitments are required for the investor to make gains from his investments. According to (Amadi 2008), acquiring land for building a hotel is relatively easier and cheaper in the USA as compared to Nigeria where exorbitant prices of landed property and high rates of lending provided by Nigerian banks discourages investors from doing business in Nigeria.

The hotel industry in Nigeria experiences challenges centred majorly on high cost of operating captive generators by means of using diesel powered generators as an alternative for ensuring constant supply electricity in hotels, this is due to the state of epileptic power supply in Nigeria. With respect to the growing rate of pollution and degradation occurring in the environment, the awareness on the need to adopt and enforce more effective measure for environmental protection is growing. Sustainable development of the natural and business environment has therefore become a crucial topic and a vital point of concern in today's world. The collaboration of policy makers, academia, industry, the general public and many other stakeholders is needed to address the issues of sustainability at all levels. Many studies have therefore shown the importance of the hotel industry in leading this process (Bohdanowicz 2005). It is therefore important for potential hotel investors to include feasibility studies of environmental impact assessment (EIA) and techno-economic analysis as regards energy efficiency and sustainability in the business plan. The knowledge gained by investors on energy and environmental challenges will reduce the negative effect of their activities on the environment. According to (Tzschentke et al. 2008), limited levels of awareness on sustainability, energy efficiency and environmental footprints amongst small business owners has constantly been a hindrance to change of harmful business activities. It is therefore hoped that this study would provide adequate knowledge on energy efficiency and sustainability in order to checkmate the activities of the Hospitality businesses and industry in Nigeria.

4 Energy Audit and Building Information

The energy consumption benchmark of a building is an indication of its energy performance. It involves analysing the energy consumption with respect to other similar buildings of same category. In order to achieve this, significant determinants that influence energy use are identified by studying the relationship between such factors and the building energy use. As a starting point, an energy audit is carried out. This is the monitoring, inspection and assessment of energy equipment and/or systems that consume energy so as to ensure that energy is properly and effiencientl utilized. Thus, energy audit is analogous to financial accounting with the mandate to record, analyze and summarize the series of transactions over a period of time. In order to achieve this, the facility management team surveys the energy records of all energy-consuming equipment and sub-systems of the facility, observes the trend of energy use in sub-systems, identifies opportunities of energy savings or areas of reducing utilisation as well as identification of such improvement. Energy Audit is an effective energy management tool that helps in the identification and implementation of strategies of achieving energy efficiency and conservation. Furthermore, not only is energy savings achieved, the equipment/system lifespan could be extended. This, in long run, translates to savings in terms of capital and reduction in emissions.

The Energy audit is a top-down scheme whose efficacy principally depends on the available resources allocated by the building management. These include:

(a) Commitment to the energy conservation and environmental protection;
(b) Expectations on the energy balance sheet to ascertain the amount that could be saved; and
(c) Target of the enhancement that such practices bring to the corporate image of the organisation involved.

The audit information is structured into a number of categories namely; building energy use, building physical characteristics, building operating characteristics, building services protocol and building indoor environment. More specifically, the main information that can be filtered from the above categorization includes:

- General features of the building such as Ground Floor Area (GFA), numbers of end-users, Building Architectural Details (BAD), building orientation, building facade, etc.;
- Technical properties of energy-consuming equipment/systems, design conditions and parameters;
- Building services design with system schematic diagrams and layout drawings showing system characteristics;
- Equipment/system operation records, including data logs of metered parameters on temperature, pressure, electric current, operational hours, etc.;
- Record of energy management opportunities (EMOs) already implemented or to be implemented;
- Record of maximum demand readings;
- Operation and maintenance (O&M) manuals with the testing and commissioning (T&C) reports; and
- Energy consumption bills of last or previous three years.

4.1 Normalized Energy Use Index

Various factors influence the gross energy consumption as earlier stated. Consequently, buildings with diverse correlating factors possess different levels of energy performance. This implies that unless facilities/buildings have exactly the same physical and climatic conditions, as well as the operational and consumer behaviours; then normalization of the energy consumption index is inevitable. This is mainly essential for the purpose of eliminating the influence of abnormal factors; thereby gives rooms for easy and appropriate comparison of such buildings with another.

It must be quickly stated that there is a paradigm shift in the operation and service provision in the industry. The current trend in the business is to engage more energy efficient approach for value added service. Thus, the hotel industry is promoting the use of robots for provision of services in the hotels; hence the study of the energy supply and demand is becoming more important so as to be able to prepare the industry in Nigeria for this nature of service.

4.1.1 Climatic Normalization of Energy Consumption

Since the space cooling or heating degree-days are essential metrics for determining the energy intensity of buildings; thus the energy use in buildings, especially residential spaces, are affected by climatic conditions. Hence there is need to provide some normalization based on the local time loading as well as year-to-year variations in climates. This will help to eliminate the influence of the climatic conditions on energy consumption. There are two basic techniques for addressing the importance of the weather and climatic effects on the energy consumption; these are the simple average method (Li et al. 2014a, b) and the degree-day approach (Xydis et al. 2009).

So, the simple average approach finds the average of the facility's energy consumption over the past years. Based on the work of Xin et al. (2012), this is easily determined as methematically expressed in Eq. 1

$$FEC_{clt} = \frac{1}{n}\sum_{k=1}^{n} FEC_k \tag{1}$$

where FEC_{clt} is the facility's climate adjustment value of the building energy consumption.

FEC_k is the facility's energy consumption in year k; and n is the number of years.

4.1.2 Degree-Day Method

The Degree-day approach is a method in which weather data is derived through computation of outside air temperature readings. So, using this approach, thus the heating degree-days (HDD) and cooling degree-days (CDD) are two variants of the degree-days methods deployed for extensive calculations of building energy consumption. Tantiwanit (2007) used cooling degree-days method to show the relationship between outdoor weather conditions and energy demanded for office buildings in Bangkok. The work showed that the cooling degree-days in any measured year was different from the given cooling degree-days in a standard year because of the differences in weather conditions of both years.

Meanwhile, Hong et al. (2014) proffered that energy use can be normalized using cooling degree-day (CDD) and heating degree-day (HDD) for the purpose of benchmarking; however it also stated that related data may be inaccurate despite being available because buildings differ in sizes and usage and may lead to incorrect conclusions for benchmarking. It was also indicated that since CDD/HDD was calculated using approximation method, it is therefore safer to avoid deeper forms of benchmarking with normalized energy use. Krese et al. (2012) on the other hand used the CDD to analyze real electrical energy consumption data and compared the analysis with another analysis estimated from a derived improved degree-day method using latent load. A metered data was consequently used to get the base temperature and calculation technique of degree-day after several combinations of both methods. There are a number of constraints which are the definition of base temperature and definition of CDD method. These were shown to be capable of affecting the application of degree-days method to cooling energy use in buildings.

MacDonald (2001) explained how heating degree-days and air-conditioning activties contribute to variation in the regressed EUI used for the conversion of a model. It was shown how heating and cooling energy tend to cancel out each other in order to get the total energy for a year. Chung et al. (2006) highlighted how the energy consuming supermarkets in Hong Kong is adjusted according to the weather conditions. It displayed how the adjustment was made based on the degree-day that occurred within the 12-month energy consumption period. Thus, leading to the development of a formula which represents the adjustment factor that shows how corresponding degree-days occurring during the period under review. was adjusted based on the average energy consumption of the previous twenty years to evaluate annual cooling degree days in Hong Kong was given by;

$$\text{Adjustment factor} = \frac{CDD_{20years}}{CDD_{supermarket}} \tag{2}$$

where $CDD_{20years}$ represented the average value of the previous 20 years for the annual cooling degree-day and $CDD_{supermarket}$ is the corresponding degree-days of 12-months in the recorded period.

In another work, Duarte and Matos (2006) characterized the heating and cooling degree-day by simple difference method as depicted in Eqs. 3 and 4.

$$HDD = \sum_{t_{start}}^{t_{end}} (\theta_{HDD_{base}} - \theta_e) \tag{3}$$

$$CDD = \sum_{t_{start}}^{t_{end}} (\theta_e - \theta_{HDD\_base}) \tag{4}$$

This is a difference between the base temperature and outdoor temperature whereby only the positive values are considered for the HDD while for the CDD, the difference is between the outdoor and base temperature. In other words, the HDD is calculated in such a way that the no values can be used for its caculations above the base temperature while for the CDD is evaluated at the values above the base values where HDD and CDD are in degree Celsius days, $\theta_{HDD_{base}}$ and θ_{HDD_base} represents base heating season temperature and base cooling season temperature both in °C, θ_e represents the outdoor temperature on a daily basis in °C, t_{start} represents starting day of heating or cooling season (day), t_{end} represents for the end day of heating or cooling season (day).

According to Chartered Institution of Building Services Engineers (CIBSE) report, the degree day data is defined as climatic data estimated from the external environment of the building (CIBSE 2006). It further explained that the mean degree hour can be used to evaluate degree day by summing up only positive hourly temperature differences and dividing by twenty-four hours. The heating degree and cooling degree day formulae were given in 5a and 5b.

$$HDD_d = \sum_{j=1}^{24} (\theta_b - \theta_{o,j})_{((\theta_b - \theta_{o,j}) > 0)} \tag{5a}$$

$$CDD_d = \sum_{j=1}^{24} (\theta_{o,j} - \theta_b)_{((\theta_b - \theta_{o,j}) > 0)} \tag{5b}$$

where D_d is the daily degree days for a single day, θ_b is the base temperature and $\theta_{o,j}$ is the outdoor temperature in the hour J.

Degree-days technique requires large amount of data and storage when compared to other methods. Generally, degree-days are estimated over suitable time interval usually over a season. In degree-day theory, the base temperature, or "balance point" of a building is the outdoor temperature which is higher than the temperature the building does not require heating. Different buildings possess different base temperatures. For any base load energy to be weather normalized, the base load energy value has to be removed from figures of energy-consumption. This is simple in theory, but very difficult in practice (Fig. 1).

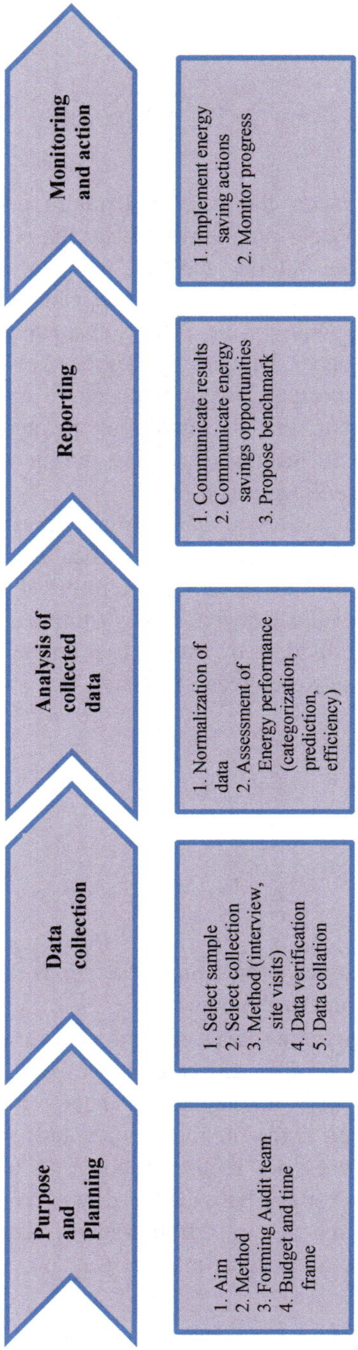

Fig. 1 Energy audit framework

5 Method

5.1 Data Collection

The only known work on the energy performance analysis of hotels in Lagos, Nigeria was carried out by adapting the statistical tools. The process was divided into three segments namely: collection of building information and energy consumption data, empirical identification of factors that influence the energy consumption pattern and evaluation of carbon-dioxide (CO_2) emission (as a product of energy consumption) by the individual hotels.

The unavailability of the energy consumption data for hotel buildings in Nigeria is a grave challenge. Simply, the access to energy consumption data in Nigeria's hotel industry is limited; which in most cases, does not exist. This thus means that energy audits are only carried out when an energy assessment/upgrade on energy-consuming facilities is necessary. As a result, data used for this study are basically from primary sources and limited in quantity; though not in quality. The foregoing inspired the adoption of the strategic quantitative and qualitative data gathering approach for this study which was obtained through walk-through energy audit, questionnaire survey, as well as personal interviews with related professionals in the industry such as the facility managers and hotel managers. Furthermore, special attention was paid to the geographical location, patronage, and star rating, hotels samples were chosen (seven in each category). Thus, the selected Hotels are located close to the airport (mainland), seaport (Lagos island) and higher educational institutions. Hence, the sample is indicative of the most visited star-rated hotels in Lagos metropolis. Questionnaires that reflect the basic building information and energy consumption inventory was developed and administered to energy experts in hotel industry, hoteliers and facility managers of the sampled hotels. The basic information gathered include: year of construction of building, gross floor area, occupancy rate, number of guest rooms, number of employees, star ratings and electric energy consumed by the selected hotels (see Tables 2 and 3). However, there were difficulties in establishing water and gas consumptions of majority of the hotel samples. For the purpose of this chapter, only the data on electricity consumption was obtained.

Meanwhile, the sampled hotels were heterogeneous in terms of size and star-rating. The statistical summary of the data collected in respect of the 28 hotels

Table 2 Summary of collected parameters

No.	Item	Included parameter
1	Building physical characteristics	Total floor area (TFA), age of building
2	Building energy use	Electricity bills
3	Building operation characteristic	Number of guest, star rating, number of workers, occupancy rate

Table 3 Summary statistics of surveyed hotels

Factor	X_i	Min.	Max.	Average	Standard deviation
TFA (m^2)	x_1	1020	64,464	20,321.61	18,892.75
Age (year)	x_2	1	72	15.25	19.66
Number of guest rooms	x_3	8	1475	289	375
Occupancy rate (%)	x_4	40	100	72.61	15.57
Guest-night sold[a]	x_5	10	412.75	116	116
Number of employee	x_6	10	654	167	178

[a]Guest-night sold is the product of the number of guest rooms and occupancy rate

is shown in Table 3. The smallest hotel has 10 guest rooms with the minimum TFA of 1020 m^2. The largest hotel houses 654 guest rooms with TFA of 64,464 m^2.

5.2 Building Energy Use Intensity

Due to the fact that there is no universal scale or standard for representing energy benchmarking of buildings it is challenging to compare the energy utilization between similar-purpose buildings. Simple measurement of energy used in specific time duration does not take into account some building characteristics such size of building, weather and climatic conditions, configuration or type of use etc. Energy Use Intensity (EUI) has been widely adopted by numerous researchers. This index provides the means to standardizing the approach such that energy use is normalized and as such can be compared for various types of buildings in the same category. EUI estimates the means of reducing total energy consumption. The EUI is expressed as a function of a building's *gross floor area* or "*footprint*". Hence, the Energy Use intensity (EUI), is the ratio of annual energy consumption to the Total Floor Area (TFA) of a building. It accounts for the variance in energy use due to difference in facility floor areas and is used for comparison of energy consumption among buildings of its peer. Essentially, the EUI specifies a building's energy use with respect to its size or other building characteristics. The EUI is also regarded as the Building Energy Performance (BEP). This is mathematically expressed as:

$$EUI = \frac{AEC}{TFA} \tag{6}$$

where *AEC* is the annual energy consumption, *TFA* is Total Floor Area.

Table 4 shows detailed information obtained from the sampled hotels across Lagos metropolis. In this case, it provides a brief review of the hotels with their corresponding annual energy consumption values. While information is thus provided on the energy been utilized by all the functional electrical appliances of the hotel in a given year, star rating, total floor area in square meter, age; signifying the number of years since inception till date, number of guest rooms, occupancy rate,

Table 4 Detailed information of the 28 sampled hotels

Hotel	Annual energy consumption (MWh)	Star rating	Total floor area (m^2)	Age (year)	Number of guest rooms	Occupancy rate (%)	Guest-night sold	Number of workers	EUI (kWh/m^2)
H1	831.1225	2	1822.986	15	20	80	16	36	455.91
H2	3217.268	2	5346.02	36	74	50	37	150	601.81
H3	2800.782	2	18,032	16	70	80	56	100	155.32
H4	5167.743	4	10,197.07	2	65	40	26	130	506.79
H5	4580.832	3	12,200.076	6	74	85	62.9	130	375.48
H6	8040.813	4	60,000	5	92	100	92	160	134.01
H7	22,092.13	5	55,196	6	64	75	48	300	400.25
H8	26,400.26	5	40,907.0375	37	654	56.53	369.7062	1475	645.37
H9	11,038.9	5	33,864	29	332	85	282.2	526	325.98
H10	1709.069	4	8670	4	85	65.1	55.335	75	197.12
H11	1196.99	4	3672	3	36	82.9	29.844	35	325.98
H12	4613.987	5	23,868	4	234	75.3	176.202	562	193.31
H13	6940.698	5	35,904	1	352	75	264	650	193.31
H14	12,520.86	5	64,464	54	635	65	412.75	1345	194.23
H15	8777.923	4	26,928	3	264	90	237.6	310	325.98
H16	828.1515	3	4284	64	42	60.2	25.284	50	193.31
H17	288.4302	2	6324	11	62	70	43.4	85	45.61
H18	558.252	2	20,460	4	120	82	98.4	163	27.29
H19	46.521	2	1020	11	10	100	10	8	45.61
H20	3923.861	3	18,212	3	199	75	149.25	87	215.45
H21	9210.17	4	28,392	72	277	70	193.9	362	324.39
H22	106.9983	2	2104	3	23	85	19.55	25	50.85

(continued)

Table 4 (continued)

Hotel	Annual energy consumption (MWh)	Star rating	Total floor area (m^2)	Age (year)	Number of guest rooms	Occupancy rate (%)	Guest-night sold	Number of workers	EUI (kWh/m^2)
H23	884.6944	3	4500	15	44	80	35.2	79	196.60
H24	1407.468	3	6405	4	70	40.1	28.07	120	219.75
H25	4721.459	4	12,567	5	142	50	71	105	375.70
H26	9287.128	5	48,042	5	471	60.3	284.013	764	193.31
H27	847.8694	3	4200	5	43	80	34.4	55	201.87
H28	3723.968	3	11,424	4	112	75.6	84.672	185	325.98

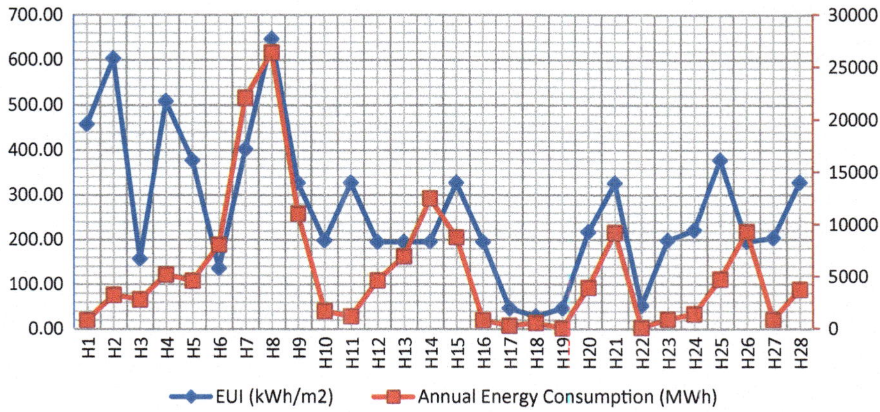

Fig. 2 Relationship between EUI and annual energy consumption

number of equivalent guest rooms; which is the product of number of guestrooms and occupancy rate, number of workers and the energy use normalised using total floor area. This thus suggests that the annual energy consumption can also be normalised using the number of guestrooms, equivalent number of guestrooms, number of workers as well as weather conditions. Generally, models are expressed in the form of Energy Use Intensity (EUI) with units in form of Kwh/m^2/year or MJ/m^2/year. Alternatively, a unit of kg CO^2/m^2/year with respect to environmental implication can be adapted. All these units make benchmarks easy to understand for professionals and non-professionals.

Thus, five possible normalisation indexes are achievable. Also, displayed in the table is the EUI for each hotel. Figure 2 shows the relationship between the annual energy consumption and the EUI for the 28 hotel samples.

5.3 *Correlation Analysis*

In order to establish the degree of association between the many influencing variables of energy use, Xin et al. (2012) and Oluseyi et al. (2016) adopted the correlation analysis approach. Unlike ordinary least square (OLS) method that fits a line that shows the relationship of the plot, the correlation analysis simply quantifies how well two variables relates with each other. Additionally, correlation analysis can only estimate the index explaining the relationship between two variables but with a regression analysis model the relationship between more than two variables can be predicted. Correlation analysis is usually used to study the degree of association between two sets of variables centered on the data retrieved either by observation or experiments (Takane et al. 2006). It identifies the most significant factors which may then be used to normalize the energy consumption (Oluseyi et al.

Table 5 Pearson correlation between EUIs and variables

Indexes	Energy consumption per unit area	Energy consumption per unit guest room	Energy consumption per equivalent guest room	Energy consumption per unit employee
Variable name	Pearson correlation value			
Total floor area	−0.069	0.418[a]	0.352	0.296
Age	−0.039	−0.094	−0.084	−0.154
Number of guest rooms	0.046	0.122	0.117	0.175
Occupancy rate	0.04	0.007	−0.178	−0.001
Number of equivalent guest rooms	−0.051	−0.123	−0.143	−0.163
Number of employee	−0.016	0.002	0.01	−0.178

2016). The range of the correlation coefficient is $|r| \leq 1$; the closer $|r|$ is to 1, the stronger the correlation. In the event that $r = 0$, no correlation exists. Meanwhile if $|r| < 0.3$, correlation is weak; whereas if $0.3 < |r| < 0.5$, then correlation is low; while if it is $0.5 < |r| < 0.8$, it is a significant correlation and; $0.8 < |r| < 1$ shows high correlation; $r = 1$ is perfect correlation (Ma 2005). For further appreciation, correlation results after normalisation using the TFA, number of guestrooms, number of equivalent guestrooms and number of employee is as shown in Table 5.

5.4 Regression Analysis

In order to establish a model for energy consumption with '*n*' data set, let the EUI and $y_1 \ldots y_n$, be the primary building characteristics variable for normalization. These factors include star rating, TFA, age of building, number of guest rooms, occupancy rate, number of workers and equivalent number of guestrooms. Regression model is developed to characterise and classify the energy use intensity determinants. The first step involves collection of potential factors that influence energy use. This may be identified during data collection. In this regards, the regression model can be directly created if the independent variables to be used are already known. Often times this is not so. Consequently, a list of possible energy use drivers is accessed for their degree of significance and correlation with the dependent variables and other primary independent variables. The linear regression technique can be used in this regard. Only statistically significant variables $y_1 \ldots y_n$, are left in the model. A best-fit multiple regression model is structured from the collected data which has the final form described in Eq. 7.

$$EUI = a + b_1 y_1 + \cdots + b_n y_n + \in \quad (7)$$

where a is the intercept; $b_1 \ldots b_n$ are the regression coefficients; while $y_1 \ldots y_n$ are the significant factors and $\in$ represents the random error.

5.5 *Prediction Using Multiple Linear Regressions*

From the information in Table 4, the data can be analysed to deduce which variables significantly affect the energy consumption intensity or otherwise. Thus, the energy consumption is selected as the dependent variable while all other variables discussed in previous sections are independent.

5.5.1 Relationship Between Annual Energy Consumption Total Floor Area

Figure 3 displays a plot of the annual energy consumption in MWh against the total floor area in square metres (m^2). From this figure, 28 data points, representing the data size, are displayed and the plot of the respective trendline is as shown in the figure. The equation depicting the trendline is also shown. It is observed that the R^2 value of the plot is 0.5883 which is moderately high. Note that the value 0.5883 implies that the total floor area can deduce the annual energy consumption by 58.83% in the hotels.

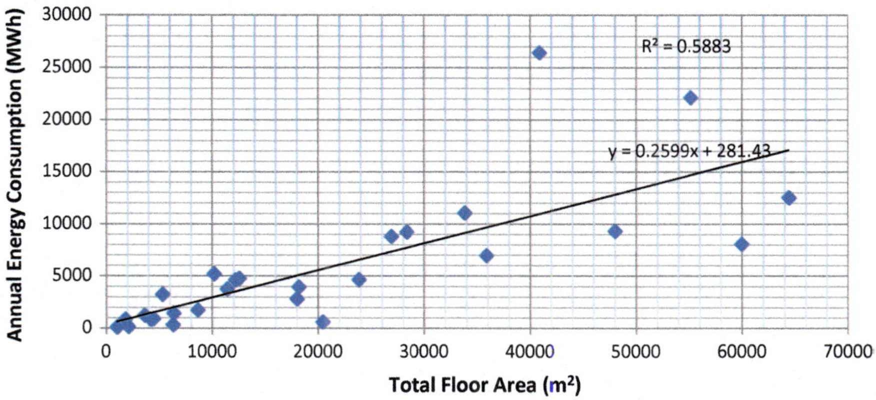

Fig. 3 Annual energy consumption against total floor area

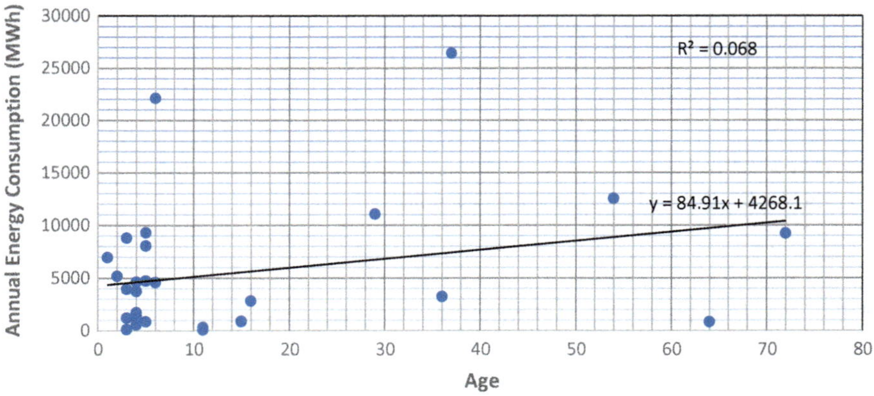

Fig. 4 Annual energy consumption against age

5.5.2 Relationship Between Annual Energy Consumption and Age of Building

Figure 4 displays a plot of the annual energy consumption in MWh against the age of each hotel building. From the plot, the equation of the straight-line graph (trendline) is obtained, likewise the R^2 value of about 0.068 which is very low. The implication of this is that the age of the hotels is rather insignificant for the prediction of the annual energy consumption by the hotels.

5.5.3 Relationship Between Annual Energy Consumption and Number of Guest Rooms

Figure 5 shows the plot of the annual energy consumption against the number of guest rooms of the hotels. The straight line (trendline) equation is shown with the corresponding R^2 value of 0.4813. The R^2 value of 0.4813 can be used to explain approximately 48.13% of the variation in energy use of the hotels which can be considered significant to the total annual energy consumption.

5.5.4 Relationship Between Annual Energy Consumption and Occupancy Rate

Figure 6 shows the plot of energy consumption against occupancy rate. Likewise, the trendline equation is also depicted in the diagram as well as the R^2 value of approximately 0.0166. The regression value (R^2) obtained from the plot is very low as compared to the other variables, suggesting that the influence of occupancy rate on the energy consumption is very low and could be said to be negligible.

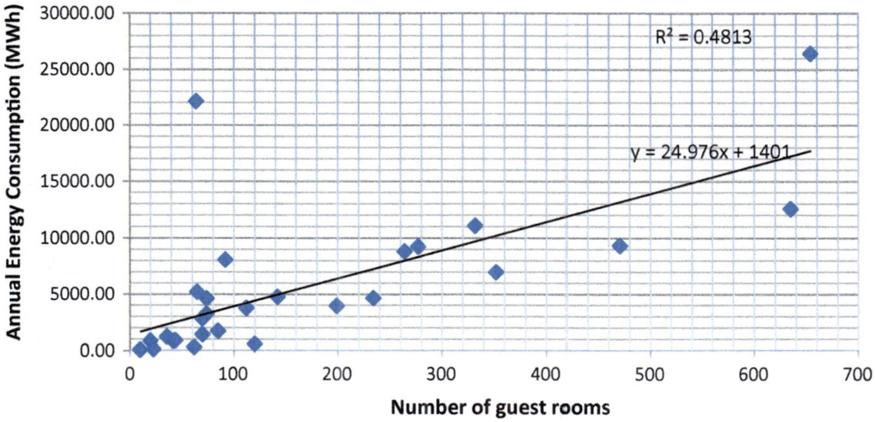

Fig. 5 Annual energy consumption against number of guest rooms

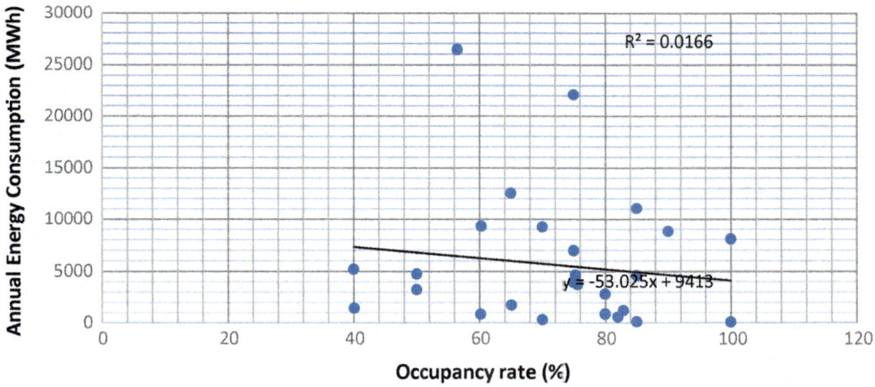

Fig. 6 Annual energy consumption against occupancy rate

5.5.5 Relationship Between Annual Energy Consumption and Number of Equivalent Guest Rooms

Figure 7 shows the variation between the annual energy consumption in MWh and the number of equivalent guest rooms. It is observed that the regression value, R^2, is approximately equal to 0.4379 signifying that about 43.79% of the annual energy consumption can be explained by the number of equivalent guest rooms of the hotels. (The number of equivalent guest rooms is the product of the occupancy rate and the number of guest rooms).

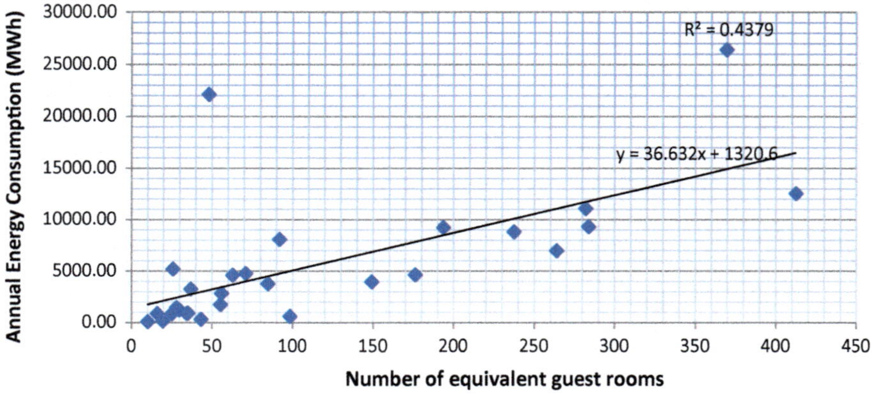

Fig. 7 Annual energy consumption against number of equivalent guest rooms

5.5.6 Relationship Between Annual Energy Consumption and Number of Workers

Figure 8 shows a plot of energy consumption against the number of employees for each hotel, the straight line (trendline) equation and the regression value (R^2 value) of approximately 0.5743. The regression value (R^2 value) of 0.5743 is quite significant to the energy use and able to explain about 57.43% of the energy consumption of the hotels.

Generally, the higher the star rating of a hotel, the more the hospitality service rendered by the hotel which thus results in more energy consumption by the hotel. This work also analyzed the effect of star rating of hotels on energy use intensity. Energy use intensities of hotels are plotted against the star rating of individual hotels (Fig. 9). It can be observed that the EUI varies across the star rating. For example, there are some 2-star rated hotels that perform comparably in terms of EUI with the 5-star rated hotels. It can thus be established that the EUI for the individual hotel buildings does not strictly correlate with their star ratings.

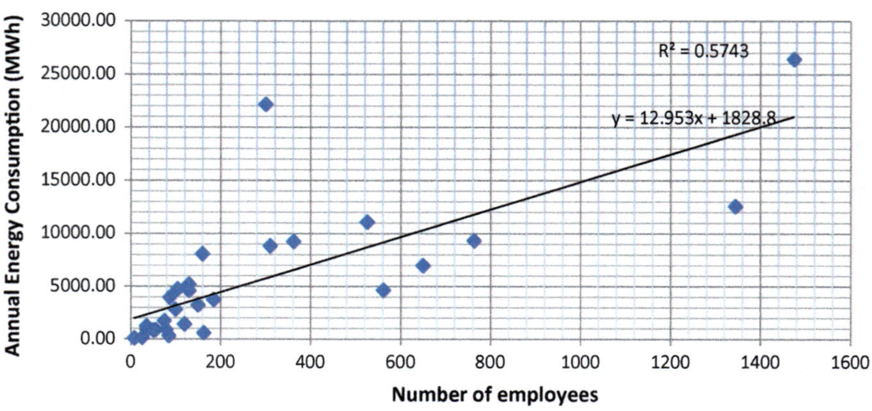

Fig. 8 Annual energy consumption against number of employees

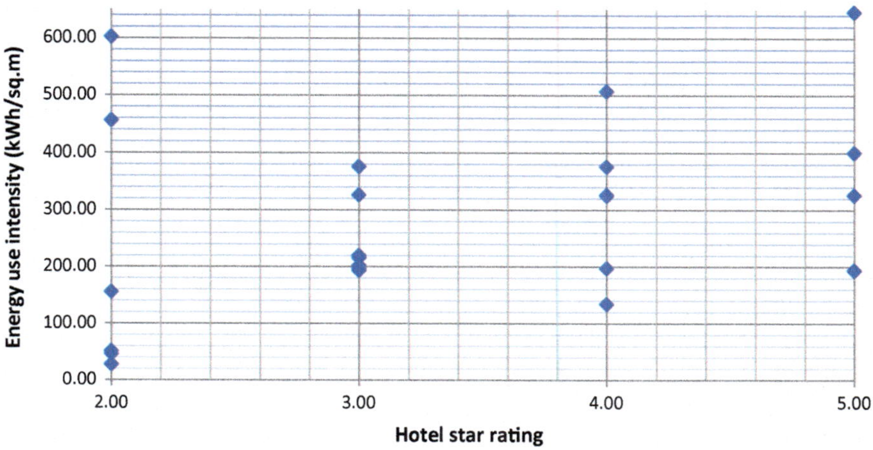

Fig. 9 Relationship between EUI and star rating of hotels

From the regression analysis, it is observed that all the independent variables have effect on the energy consumption pattern but in varying degrees and level. This would therefore lead to considering some variables as significant while others are treated as insignificant to the dependent variable, energy consumption, based on the degree of the regression value. Table 6 gives a summary of the variation of each independent variable with respect to the dependent variable. Considering the R^2 relationship of each independent variable with respect to the dependent variable, the following significant variables could be selected for energy prediction:

i. Total floor area (x_1)
ii. Number of guestrooms (x_3)
iii. Guest-night sold (x_5)
iv. Number of employee (x_6).

The variables which are significant could be said to be so on the basis of the fact that:

i. The total floor area actually reflects the possible available space for construction of rooms, offices, lobby, installation of electrical appliances like lighting appliances etc.; which would require more energy for usage.
ii. The number of guestrooms reflects the available space for lodging and relaxation. In other words, hotels with more guest rooms will usually require more energy supply for consumption.
iii. The number of workers actually reflects the degree of activities in a hotel. In essence, the high number of workers in a hotel signifies that there is a lot of day to day running business activities in such hotel.

Based on the significant variables identified, an equation could be derived for predicting the energy consumption in the hotels using Multiple Linear Regression (MLR) method. The result of this (i.e. multiple linear regression) analysis leads to the proposed development of (gives) Eq. 8. The R^2 value for the model is

Table 6 Regression between dependent variable (energy consumption) and independent variable

		Total floor area (m^2)	Age (years)	Number of guestrooms	Occupancy rate (%)	Number of equivalent guestrooms	Number of workers
Annual energy consumption (MWh)	R^2	0.5883	0.068	0.4813	0.0166	0.4379	0.5743
	R	0.7670	0.2607	0.6937	0.1288	0.6617	0.7578

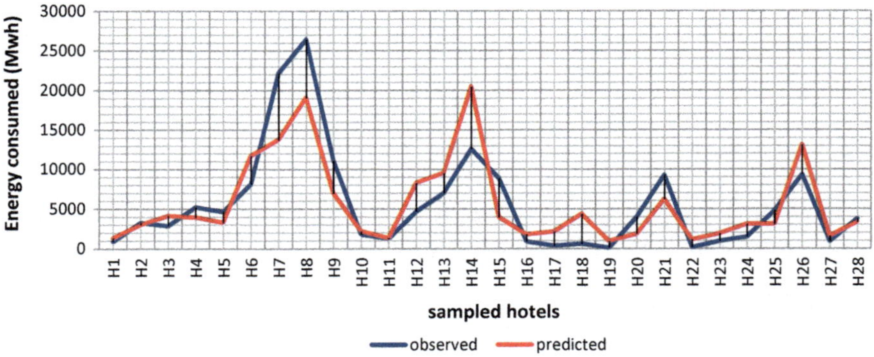

Fig. 10 Comparison of predicted and observed energy consumption of sampled hotels

approximately 0.71 showing that there is a strong relationship between the selected dependent variable and the independent variable. Using Eq. 3 and the collected data, a plot showing the comparison between the predicted and observed data is given in Fig. 10.

$$Y = 847.9101 + 0.18422x_1 + 4.354637x_3 - 29.34223x_5 + 12.63989x_6 \quad (8)$$

5.6 Other Benchmarking Methods

According to Xin et al. (2012), various statistical methods can also be used to investigate the energy consumption benchmark of hotel buildings. Such methods include the mean index of total energy consumption (MITEC), mean of EUIs, quadratic average method, median, percentile method as well as the mode. Apart from these statistical methods, some other methods involve implementation of metaheuristic estimating approach for support vector machines (SVM), artificial neural network (ANN), non-parametric Data envelopment Analysis (DEA), etc., have found applications in energy benchmarking.

5.6.1 Mean Index of Total Energy Consumption

The mean index of total energy consumption (MITEC) is a performance index that defines the building energy consumption performance as the ratio of the climatic adjusted building energy consumption for a year to the summation of the building area, number of guestrooms and any other influencing significant factor. The MITEC is the representative of the gross level of energy consumption in a zone. A major shortcoming is it inability to accurately reflect the energy use level of the single building. This can be evaluated using Eq. 9; where E_i is energy consumption and A_i is the building area.

$$MITEC = \frac{\sum_{i=1}^{n} E_i}{\sum_{i}^{n} A_i} \tag{9}$$

5.6.2 Mean of Energy Use Intensity (EUI)

Energy use intensity can also be referred to as the Energy Use Index. Its mathematical expression is as shown in Eq. 6. On the other hand, the Mean of EUI estimates the average of the normalized energy use indexes of the sample size. This can be used to establish a consumption quota (Xin et al. 2012). It is thus calculated using Eq. 10.

$$ECQ_{MEUI} = \frac{1}{n}\sum_{i=1}^{n} EUI_i \tag{10}$$

where ECQ_{MEUI} is the mean of EUI, EUI_i is the Energy Use Index and n is the number of samples. The energy consumption benchmark is based on the mean of EUIs and it gives an expression to the average level of building consumption and could be satisfied generally by buildings. Meanwhile it is not a reflection of the extent of major building energy consumption (Xin et al. 2012).

5.6.3 Quadratic Average Method

Quadratic average method is a method used for analysing the energy assumption which has been normalized whose variables include the average of the normalized Energy Use Index (EUI) and the average of the Energy Use Index which are less than the overall normalized Energy Use Index average. Following the acquisition of these two variables, a new average of the two variables is then found which helps gives the quadratic average method energy consumption. The process stated can be represented mathematically as shown in the Eq. 11.

$$ECQ_{QAM} = \frac{\frac{1}{n}\sum_{i=1}^{n} EUI_i + \frac{1}{m}\sum_{j=1}^{m} EUI_j}{2} \quad (11)$$

where ECQ_{QAM} is the Energy Consumption Quota derived through the Quadratic Average Method

EUI_i is the normalized Energy Use index of the facilities

EUI_j is the normalized Energy Use index which are less than the overall average of the normalized Energy Use Index.

The above process stated above shows that the energy consumption derived from the quadratic average method is between the average and advanced level.

5.6.4 Median

Median gotten from the word "middle" in statistics is defined to be the value which separates a set of data into two halves/parts; the higher part and lower part. It is deduced by first re-arranging the sample data in terms of ascending or descending order after which the middle value (median) of the sample data is selected. In calculating the energy consumption, median is also an available option considering the middle value of the normalized energy use index of the facility data. The middle value deduced is not affected by the highest and lowest value.

5.6.5 Percentile Method

In statistics, a percentile is a measure used to show the level or value in which a certain amount or percentage of dataset fall below. With respect to energy consumption in buildings, it is used to indicate a value below which the Energy Use Index of certain percentage of the facilities are lesser (Xin et al. 2012).

5.6.6 Mode

In analysing the energy consumption of a facility, the mode is a statistical method considered based on its normalized energy use index (Xin et al. 2012). According to statistics, the mode in a given set of data of distribution is the value that appears more often or more frequently. This is calculated by first rearranging the data sample in groups and finding the value or group which appears more often, having the highest frequency. Afterwards the mode can be found using the Eq. 12.

$$ECQ(mode) = L + \frac{\Delta_1}{\Delta_1 + \Delta_2} \quad (12)$$

where

L	is the lower limit of the modal group
Δ_1	is the difference between the frequency of the modal group and the group just below it (previous group)
Δ_2	is the difference between the frequency of the modal group and the group just after it (next group)
C	is the class interval of the groups.
ECQ_{mode}	is the energy assumption Quota of the facility calculated.

Therefore, the energy consumption derived through the mode shows the energy consumption which most facilities possess and is relevant to circumstances in a sample with the obvious control tendencies.

5.6.7 Support Vector Machine (SVM)

Support vector machine is a type of black box method that uses data fitting techniques requiring an initial statistical model with a training data. It's a type of kernel method, and belongs to an algorithm class for analyzing patterns. It has an advantage in its performance due to its higher accuracy in prediction. A mapping function for its input-output is usually obtained from a group of training data. Support vector machine can be either be support vector classification or support vector regression. For the support vector classification, the mapping function used transforms the input data using the kernel functions of a higher (may be infinite) dimensional space. The model produced therefore depends only on a division of the training data close to the boundaries of the class. Likewise, for the support vector regression, the model produced does not take into consideration training data close enough to the prediction of the **model** (Wang 2005). In particular, for the support vector regression, assuming we have an input vector A and an equivalent output vector B, SVM correlates A and B with the Eq. 13.

$$B = C.\emptyset(A) + k \tag{13}$$

where c and k are constants

$\varnothing$ is a mapping function which transform A to a higher dimensional space.

SVM has 2 major advantage over other methods. Firstly, it does not consider training examples with errors lesser than predefined constant, secondly, besides minimizing training error, it minimizes the structure risk defined by norm of C (Li et al. 2014a, b).

Zhao and Magoulès (2012) investigated the influence of building characteristics on the performance of a model by training the SVM model on multiple buildings heating load and tested it on the load data of an entirely new building. Owing to the fact that the training data size harnessed from multiple buildings will be very large, the training of the model was very slow. Therefore, in order to increase the speed of the process, parallel SVM was applied (Zhao and Magoulès 2012). Li et al. (2010) forecasted the electricity consumption of buildings in a year by applying Support

Vector Machine and Neural networks. Among all the forecasting models, Support Vector Machine was observed to have the best performance.

5.6.8 Data Envelopment Analysis

Data Envelopment Analysis (DEA) is a benchmarking technique used for examining and analysing data in public sector and non-profit organizations. DEA is known for its potentials in employing unique strategies for the improvement of services. The method of DEA is completely dependent on available data and makes no assumption on the form of data distribution when measuring performance. The performance of hotel buildings can be analysed using DEA multiple input and multiple output data for hotel buildings are available. DEA makes use of the method of linear programming for the transformation of multiple-input hotel variables (like outdoor temperature, use of electricity, food and covers for beverages, use of town gas and use of water) into multiple-output hotel variables (likes guests available in rooms, type of food, number of nights spent in rooms and covers for beverage). DEA is a very important tool which decision makers and hotel managers in the hotel industry can use for the promotion of energy use efficiency in the hotel industry (International Tourism Partnership 2014).

Some research works have been dedicated to the used of DEA in benchmarking energy use in buildings. According to a DEA model developed by Hui and Wan (2013), the relative terms of efficiency for decision-making units in a hotel identified are the outlets for beverages and the type of food served in hotels (Hui and Wan 2013). The basic concept of DEA involves the generalization of the measure of the efficiency of the single-input/single-output data in relation to the multiple-input/multiple-output data through the construction of an efficiency score in relative terms. This involves the measure of a single virtual output compared to a single virtual input. The efficiency score is defined in Eq. 14.

$$Efficiency\,Score = \frac{sum\,of\,outputs\,in\,weighted\,terms}{sum\,of\,inputs\,in\,weighted\,terms} \tag{14}$$

DEA was also used for the identification of factors influencing the operational efficiency of 58 hotels in the sultanate of Oman (Oukil and Al-Zidi 2017). Their methodology involved a two stage framework with the first stage making use of DEA for the estimation of the efficiency scores of the various hotels while the second stage involved the use of the analysis of econometrics for the discernment of any possible correlation that exists between factors relating to context (these include; environmental and non-discretionary variables which can influence efficiency but are not easy to control) and technical efficiency. The multiple-input variables for analysis include the number of beds, salary of employees, number of rooms and number of employees while the multiple-output variables for analysis include the rate of occupancy, number of nights, total cost of revenue and number of guests. Size of hotel, star rating and culture were identified as the factors having the most influence on energy. In a similar study, DEA was applied to develop a set of benchmark that will meaningfully control best practices for success in the hotel industry using 39 hotels in Korea which are recognized both regionally and internationally (Min et al. 2008). The authors selected four variables as input for their model which include the assets, cost of landed property, operating expenses and building capacity. The outputs chosen include the type of food/beverage, number of rooms and number of conventions/shows. The cost of landed property and building capacity were observed to have the most influence on the financial performance of the hotels.

In another study, Yoon and Park used DEA on two office buildings to show the performance-oriented, objective and rational qualities of DEA using multiple-input and multiple output variables (Yoon and Park 2017). Thermal energy consumption [Energy use intensity (EUI)] was the only input considered while occupant density, operation time, indoor thermal comfort and indoor air quality (in terms of carbon emissions) were considered as outputs. Carbon emissions and Energy use intensity were observed to have the most influence on performance. A study that analysed the efficiency levels of 28 hotels in Portugal with four and five star ratings using Data Envelopment Analysis (DEA) has been presented (Pereira Oliveira et al. 2015). The authors selected input and output variables related to quantitative and monetary units. The input variables included number of rooms, number of employees, economic costs, capital costs and all other costs other than economic and capital costs. The target output variables included total revenue obtained from running the hotel. Results show that, the number of employees and the number of rooms were observed to have the most influence on performance. Manasakis et al. analyse and compare the efficiency levels of independent-operating hotels and hotels operating under a brand name in the Greek Island of Crete using DEA (Manasakis et al. 2013). The inputs chosen were based on their ability to influence specific managerial goals and they included the number of employees, the number of beds and the total operational cost of a hotel. The outputs were chosen based on their ability to reflect managerial goals and objectives and they included total revenues and total number of nights spent in rooms by guests. Total revenues and total number of nights spent in rooms by guests were observed to have the most influence on performance.

A study that analysed the efficiency of hotels in Croatia to check if differences existed in the efficiency of hotels having different sizes and quality has been presented (Poldrugovac et al. 2016). The inputs considered were energy expenses, food and beverage expenses, room expenses, labour expenses and other expenses associated with other services while the outputs selected were the total revenue and the rate of occupancy. Expenses, revenue and size of hotel were reported to have the most impact on performance of hotels. The use of DEA to appraise energy efficiency level in selected Malaysian hotel has been studies (Yen and Othman 2011). The inputs selected were the number of available room nights, number of full-time employees, book value of hotel property, total operating costs (which includes employee salaries, room costs, utilities, maintenance fees and other relevant operating costs, food and beverage costs, and other expenses) while the outputs selected included number of nights which rooms were occupied, number of guests, average rate of occupancy, total operating revenues, food and beverage revenues, and other revenues. According to the results, the average rate of occupancy was seen to have the most influence on performance of hotels. Lung-Tan analysed and examined the changes in efficiency and performance of international hotels in Taipei, Taiwan using DEA (Lung-Tan 2015). The inputs used for the study included number of rooms, number of staff in charge of taking care of the rooms, number of staff working in the administration department. Others include; number of staff working in the food and beverages department, number of staff working in other departments. The outputs selected for the study included the revenue from occupation of rooms and total revenue. Number of rooms and total revenue were observed to have the most influence on performance of hotels.

Morey and Dittman (1995) used DEA to develop a benchmark model which assesses the efficiency level of hotel managers using 54 hotels in the United States (U.S.) (Morey and Dittman 1995). The following were chosen as input variables included; the expenses for major room activities, other room-related expenses, cost of energy, expenses for property, operation and maintenance, salaries for variable advertising, fixed advertising expenses, salaries for administrative and general staff, varying advertising expenses and other administrative and general expenses. The selected output included total room revenue, facilities satisfaction index and services satisfaction index. expenses and services satisfaction index were observed to have the most influence on performance of hotels. DEA tool was used to estimate the technical and allocative efficiencies in 48 U.S. hotels (Anderson et al. 2000). The selected inputs included the number of full-time equivalent employees, the number of rooms, total gamming-related expenses, total food and beverage expenses, and other hotel-related expenses. The selected outputs included total hotel activity revenue and other related revenue. Expenses and number of rooms were observed to have the most influence on performance of hotels. Furthermore, Hwang and Chang used DEA to measure and assess the managerial performance and efficiency of 45 international hotels in Taiwan (Hwang and Chang 2003). The selected inputs included the number of full-time employees, number of guest rooms, total area of meal department and operating expenses while the selected outputs included room revenues, Food and beverages revenue, and other revenue. The number of employees and operating expenses were observed to have the most influence on performance of hotels.

In another study, DEA was applied to model issues relating to efficiency and effectiveness using a 49-unit Asia-Pacific hotel chain (Keh et al. 2005). The inputs selected included number of rooms and total expenses while the output chosen included food and beverage revenue, and revenues accrued from the occupation of rooms. The number of rooms was observed to have the most influence on performance of hotels. A cross-efficiency DEA was adopted to develop a model which examines the productivity and efficiency of star-rated hotels in 31 Chinese provinces (Tsai 2009). The selected input included the number of fixed assets, number of hotels and number of employees receiving training while the selected outputs included total revenues and percentage of occupancy. The number of hotels and percentage of occupancy were observed to have the most influence on performance of hotels. In a similar manner, a study which applied DEA to design a model for the assessment of the operational performance of twenty-three tourist hotels in Taipei, Taiwan has been presented (Wu and Song 2011). The selected inputs were the total number of employees, food and beverage capacity, and total operating cost while the chosen outputs were guest room revenue, food and beverage revenue, and other revenue. Total number of employees and total operating cost were observed to have the most influence on performance of hotels. Table 7 gives a summary of the works carried out by various authors on input and output variables for analysis of energy use in hotel buildings using DEA.

Table 7 gives a summary of the works carried out by various authors on input and output variables for analysis of energy use in hotel buildings using DEA.

5.6.9 Commercially Available Software

Some commercial software for energy footprint analysis and benchmarking also exist. There are advantages which energy benchmarking software provide which makes them more superior to manual benchmarking methods. In software benchmarking, the benchmarking process takes less time to complete and no prior knowledge on energy studies are needed to understand how to use the software for generating and interpreting reports as information on how to go about it is provided (Maxwell and Forselius 2000). Some of the commercially available software for energy footprint analysis include portfolio manager, wegowise, EnergyIQ, EUI calculator, DOE-2, Energy plus, TRNSYS etc.

5.6.10 Results from Statistical Methods

In this present work, the aforementioned statistical tools are applied to the collected hotel data. The summary of the results of the achievements offered by these statistical methods as applied to the sampled hotels is as displayed in Table 7.

From Table 8, it could be observed that out of these statistical methods, the energy consumption benchmark using the mode is minimum, while the 75th percentile returned the highest value. Meanwhile, the range of benchmark option is therefore 45.61–338.35 kWh/m^2, whose percentage range is 2.90–75%. The benchmark range only cover up to 75% of the sample size and 25% of hotel buildings with energy consumption being higher than the maximum 338.35 kWh/m^2 benchmark. Three out of the energy consumption benchmark results from the statistical methods were below the average of the EUI methods used in this research (see Fig. 11). Consequently, to demonstrate the advantage of the benchmarking, the energy consumption benchmark of two-, three-, four- and five-star hotel buildings in Lagos metropolis is recommended to be an interval in the initial implementation phase, which is the range of estimated by all the statistical methods mentioned in Table 7, and then specified to be 45.61–338.35 kWh/m^2. Therefore, the

Table 7 Published works on analysis of hotel buildings using data envelopment analysis (DEA)

References	Input variables	Output variables	Variables that drive performance (factors of influence)
Hui and Wan (2013)	Outdoor temperature, use of electricity, food and covers for beverages, use of town gas and use of water	Guests available in rooms, type of food, number of nights spent in rooms and covers for beverage	Beverages and the type of food served
Oukil and Al-Zidi (2017)	Number of beds, salary of employees, number of rooms and number of employees	Rate of occupancy, number of nights, total cost of revenue and number of guests	Size of hotel, star rating and cultural attractions
Min et al. (2008)	Assets, cost of landed property, operating expenses and building capacity	Type of food/beverage, number of rooms and number of conventions/shows	Cost of landed property and building capacity
Yoon and Park (2017)	Thermal energy consumption [energy usage intensity (EUI)]	Occupant density, operation time, indoor thermal comfort and indoor air quality (in terms of carbon emissions)	Carbon emissions and energy usage intensity
Pereira Oliveira et al. (2015)	Number of rooms, number of employees, economic costs, capital costs and all other costs other than economic and capital costs	Total revenue obtained from running the hotel	Number of employees and the number of rooms
Manasakis et al. (2013)	The number of employees, the number of beds and the total operational cost of a hotel	Total revenues and total number of nights spent in rooms by guests	Total revenues and total number of nights spent in rooms by guests
Poldrugovac et al. (2016)	Energy expenses, food and beverage expenses, room expenses, labour expenses and other expenses associated with other services	The total revenue and the rate of occupancy	Expenses, revenue and size of hotel
Yen and Othman (2011)	Number of available room nights, number of full-time employees, book value of hotel property, total operating costs which includes employee salaries, room costs, utilities, maintenance fees and other relevant operating costs, food and beverage costs, and other expenses	Number of nights which rooms were occupied, number of guests, average rate of occupancy, total operating revenues, food and beverage revenues, and other revenues	The average rate of occupancy
Lung-Tan (2015)	Number of rooms, number of staff in charge of taking care of the rooms, number of staff working in the administration department, number of staff	The revenue from occupation of rooms and total revenue	Number of rooms and total revenue

(continued)

Table 7 (continued)

References	Input variables	Output variables	Variables that drive performance (factors of influence)
	working in the food and beverages department, number of staff working in other departments		
Morey and Dittman (1995)	Expenses for major room activities, other room-related expenses, cost of energy, expenses for property, operation and maintenance (POM), salaries for variable advertising, Fixed advertising expenses, Salaries for administrative and general staff, Varying advertising expenses and other administrative and general expenses	Total room revenue, facilities satisfaction index and services satisfaction index	Expenses and services satisfaction index
Anderson et al. (2000)	The number of full-time equivalent employees, the number of rooms, total gamming-related expenses, total food and beverage expenses, and other hotel-related expenses	Total hotel activity revenue and other related revenue	Expenses and number of rooms
Hwang and Chang (2003)	The number of full-time employees, number of guest rooms, total area of meal department and Operating expenses	Room revenues, food and beverages revenue, and other revenue	Number of employees and operating expenses
Keh et al. (2005)	Number of rooms and total expenses	Food and beverage revenue, and revenues accrued from the occupation of rooms	The number of rooms
Tsai (2009)	The number of fixed assets, number of hotels and number of employees receiving training	Total revenues and percentage of occupancy	Number of hotels and percentage of occupancy
Wu and Song (2011)	The total number of employees, food and beverage (F&B) capacity, and total operating cost	Guest room revenue, food and beverage revenue, and other revenue	Total number of employees and total operating cost

Table 8 Summarized results of statistical methods

Statistical method	EUI (kWh/m^2)	Percentile (%)
MITEC	273.75	58.80
Mean of EUIs	265.95	55.80
Quadratic average	209.76	47.00
Median	208.66	44.10
Mode	45.61	2.90
75th percentile	338.35	75.00
60th percentile	324.71	60.00

energy quota could be expressed mathematically as $45.61\,\text{Wh/m}^2 \leq \text{EUI} \leq 338.35\,\text{kWh/m}^2$ (Fig. 12). Hence the hotels, whose EUI value is less than the lower range value of the benchmark interval, are regarded as buildings with low energy consumption. In contrast, the hotel buildings whose EUI surpasses the upper limit or value are defined to be the high energy-consuming hotels. More importantly, the hotel buildings whose energy consumption values are within the above stated interval of values consume energy moderately.

5.7 Applications of Clustering Method in Hotel Classification

Conventionally, building energy consumption classification is established using the cumulative frequency distribution of building EUI. Nonetheless, it is found to be unreliable; class boundaries are often loosely determined arbitrarily, thus the unbalanced class ranges can bring about a few problems or challenges. Hence, the proposed new method of hotel energy classification is established using hybridised clustering techniques. Clustering is a process which achieves the aim of data mining as it breaks down data sets into meaningful sub-classes called clusters in order to create a better understanding of the structure of the data set. Pieri et al. (2015) deployed the clustering techniques to classify (or cluster) the energy consumption of 45 hotel buildings into high, medium and low using the heating, cooling and energy consumption data of the various hotels. The study then used a component defined as the absolute value of the hotel energy saving potential to analyse the best-case scenario of each cluster with the aim of recommending best practices, setting goals and defining potentials for energy conservation so as to achieve energy efficiency and sustainable development in hotels. Xuchao (2007) used fuzzy algorithm c-means clustering technique to classify the energy performance of hotel buildings in Singapore using the data obtained from the normalised energy use intensities of all sampled hotels to take into account the energy determinants of the building. In the aforementioned study; the results obtained was compared with the standard existing results obtained from the commonly used method of classification traditionally based on cumulative frequency distribution of building energy

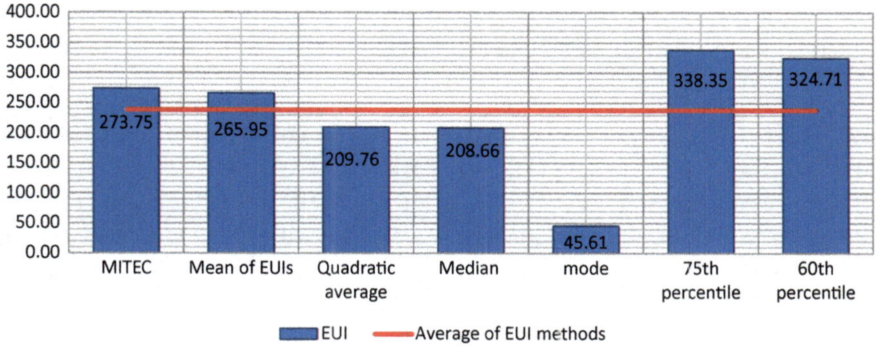

Fig. 11 Benchmark methods comparison

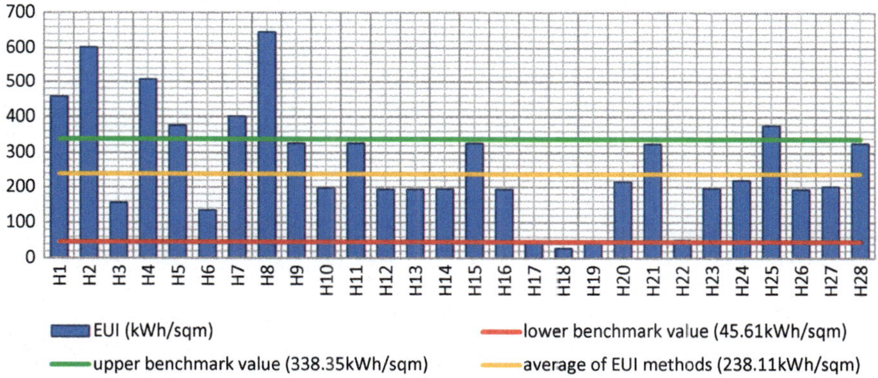

Fig. 12 Energy consumption quota interval of two, three, four and five star hotel buildings in Lagos

intensity. The work then used the results to make inferences with regards to energy efficiency projections of the hotel buildings. Furthermore, Another study used an iterative K-means algorithm clustering technique to cluster the electrical and thermal energy consumption of hotels in Greece (Farrou 2013). This approach was exploited to define the 'average/centroid' value of each cluster representing the 'typical' value of that cluster. Thus, the classification made use of the MATLAB software to showcase the essentiality of energy use development protocol. In this case, the classification of the hotels was done for 90 hotels of which only 30 hotels consume oil for waterworks and space heating. Meanwhile, another 49 hotels with annual operation of which 12 hotels consume oil for space heating and domestic hot water service provision, while there are 41 hotels with seasonal operation of which 18 hotels consume oil for domestic hot water services. The variation of energy consumption between and within clusters was very considerable; thus, indicating that it might be more suitable to use. Sun et al. used clustering analysis to create

Table 9 Cluster analysis results

Hotels	Star rating	Electric use intensity (MWh/m^2)	Class	Hotels	Star rating	Electric use intensity (MWh/m^2)	Class
H1	2.00	455.91	Medium	H15	4.00	506.79	Medium
H2	2.00	601.81	Medium	H16	4.00	134.01	Low
H3	2.00	155.32	Low	H17	4.00	197.12	Low
H4	2.00	45.61	High	H18	4.00	325.98	Medium
H5	2.00	27.29	High	H19	4.00	325.98	Medium
H6	2.00	45.61	High	H20	4.00	324.39	Medium
H7	2.00	50.85	High	H21	4.00	375.70	Medium
H8	3.00	375.48	Medium	H22	5.00	400.25	Medium
H9	3.00	193.31	Low	H23	5.00	645.37	Medium
H10	3.00	215.45	Low	H24	5.00	325.98	Medium
H11	3.00	196.60	Low	H25	5.00	193.31	Low
H12	3.00	219.75	Low	H26	5.00	193.31	Low
H13	3.00	201.87	Low	H27	5.00	194.23	Low
H14	3.00	325.98	Medium	H28	5.00	193.31	Low

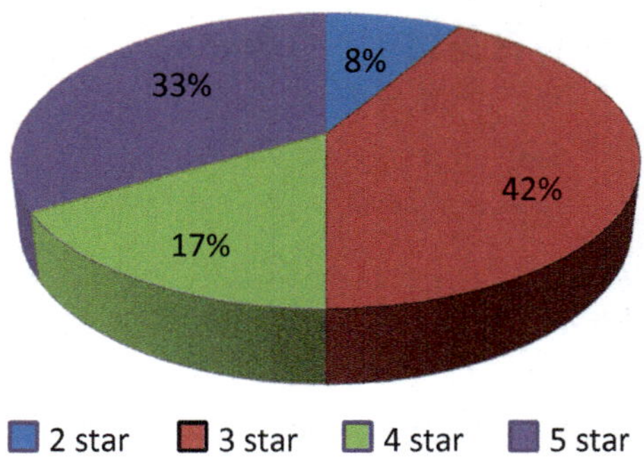

Fig. 13 Contribution of low energy consumption hotels by star rating

energy performance classes so as to identify typical buildings under tropical climatic conditions (Sun et al. 2006). These classes were based on benchmarks established using the method of normalisation and determination of performance indicators. The energy performance classes were then analysed for possible successes and failures. Thus, technically feasible energy conservation measures and energy savings potentials were then explored from the analysis. Using the Stata package version 12, the Cluster analysis (CA) was implemented. The CA implementation takes the advantage of the K-median clustering method (see Table 9).

From the results of the analysis, four hotels buildings were identified as high energy consuming. All of these are 2-star rated. There are other 12 hotels which are identified as medium and low energy intensive. As can be seen in Fig. 13, the 3-star

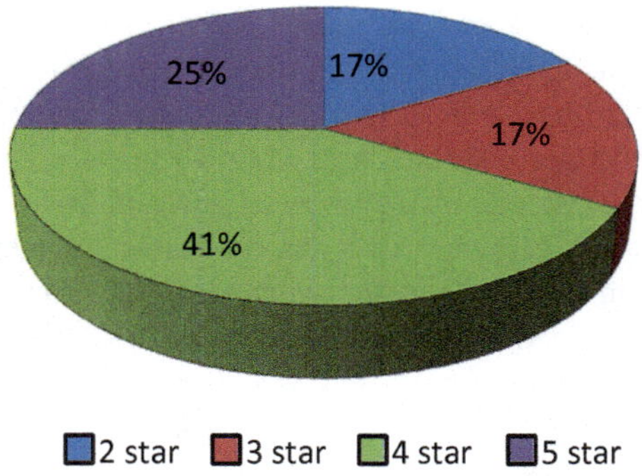

Fig. 14 Contributions of medium energy consumption hotels by star rating

hotels have the largest share of the low energy consuming facilities with a value of 42% while the 2-star rated hotels contributed the least with 8% (which is the lowest). The 4-star hotels contributed 42% of the energy consumed by the medium energy-consuming hotels buildings, while the 3-star contributed 17%, 2-star contributed 17% and 5 star contributed 25% (Fig. 14).

5.8 *Outcome of Energy Benchmarking in the Sampled Hotels*

The benchmarking protocol is established on five prongs namely; energy performance, financial performance, gaps, opportunities and implementation. While the first item has been explained fully in the preceding sections, the others can easily be obtained from the available information as captured above.

In the case of the financial performance in the hotel industry; the cost of energy as a derivative of the financial performance is quite enormous since energy consumption is a major input in the budgetary allocation for the industry's effectiveness. According to Figs. 13 and 14, the 3-star hotels have the largest share of the low energy consuming facilities which directly translates to high cost of operation due to the amount of energy consumed while the 2-star rated hotels contributed the least which follows that the energy cost would be least. So also the other categories would consume energy with the attendant cost being depicted by the efficiency of energy consumption. This would be very useful in budgetary allocation for both energy consumption and reduction in energy consumption through the installation of energy efficient appliances which in the long run improve the financial performance of the industry.

The benchmarking thus provides opportunities for the industry to pay attention to critical sources of energy waste, so as to reduce the attendant emissions. With regard to Table 9; the opportunities provided by retrofitting of the hotels with energy efficient appliances cannot be overemphasized. This includes the use of occupancy sensor facilities, automated guest rooms and multi-purpose halls as it can be seen from the case of the 4-star and 5-star hotels as the great contributors to the improved energy utilization and efficiency. But with the introduction of robotics to the hotel service provision, the approach to energy consumption would need further research to ensure that this will facilitate activities especially in the service areas such as the kitchens and laundries.

Meanwhile, the gaps noticeable in the energy benchmarking in the hotel industry may not be totally overcome if the energy demand and supply is not effectively bridged by provision of alternative energy for the industry. So the provision of hybrid energy system with great focus area on solar-powered hotels should begin to occupy the interest of the stakeholders. Though the initial cost of implementation may be high but the payback would hugely influence the profit making of the investors in a short while. Because a similar work has identified that that the payback period and life cycle cost analysis is the best that can happen to this industry.

Thus lastly, the benchmarking of the industry is highly encouraged to promote the implementation of both alternative energy supplies. The adoption of energy policy for the hotel industry in the country is very essential. Due to the carbon accounting procedure herein deducted from the energy utilization; then the carbon capture and sequestration tools need to be installed in the hotels to capture the carbon emission so as to ensure that it is not released into the environment.

5.9 Evaluation of carbon-dioxide (CO_2) Emissions in the Sampled Hotels

It is necessary to assess the emissions vis-à-vis the energy consumption values of the various hotel ratings. This is because the Greenhouse gases (GHG) trap heat in the atmosphere. Consequently it causes global warming and climate change; this has been established to influence the emission from fossil-fuel electricity generators that can contribute to the quantum of the noxious gases in the atmosphere. Meanwhile, one of the major components of greenhouse gases is carbon-dioxide (CO_2). This section provides information on CO_2 emissions from hotel sector in Lagos state, Nigeria as it relates to the expected quantity of energy consumption. The CO_2 emission from the sampled hotel can be estimated using Eq. 15.

$$Annual\,\mathrm{CO_2}emission = EU \times CEC \tag{15}$$

where EU represents the amount of electricity used by the hotel in kilowatt hours (kWh) and CEC is the CO_2 emission coefficient of the electricity used.

Due to the low reliability of the power supply from the grid, the management of most hotel buildings in Nigeria set budgetary fund aside for the diesel-driven captive electricity generation. These generators usually run for 24 h per day. For the purpose of this chapter, it was assumed that diesel generator was the only source of electricity in the hotel sector. Hence the metric tons of CO_2 emitted by the hotels was calculated using Eq. 16

$$Metric\,tons\ of\,\mathrm{CO_2}\,per\,year = AEC \times FC \times E \times CF \qquad (16)$$

where AEC represents the annual energy consumption (kWh), FC is the fuel consumption (g/kWh) of the generator used which was retrieved from the specification sheet of the generator, E is the CO_2 emission of diesel which is 22.37 lbs CO_2/gallon, CF is the conversion factor from lbs to metric tons (Oluseyi et al. 2016). Figure 15 shows the annual metric tons of CO_2 emissions from the 28 sample hotels and the individual energy use intensity in kWh/m^2. It can be observed that there is no direct relationship between them. For example, hotel 19 with the highest metric tons of CO_2 emission has a lower EUI when compared with hotel 28 which has the highest EUI while it recorded much a lower metric tonnes of CO_2 emission. It is noteworthy that the CO_2 emission varies from 8779.93 metric tonnes to 55,718,702.47 93 metric tonnes. This implies that there would be a great notch high increase in the atmospheric temperature. The impact of this on the atmosphere would thus contribute to the global warming due to the discharge of the greenhouse gas into the atmosphere. This thus means that a very serious step should be taken to overcome the concentration of CO_2 in the atmosphere.

There are some shortcomings of this study which include the fact that the concentration of the CO_2 emitted to the atmosphere is not addressed in this investigation due to lack of adequate facilities such as the carbon dioxide meters. So also, the inability of the researchers to gather so much data from large spectrum of hotels due to the fact that the hotel managers have no faith/understanding in the energy management of the hotels in the industry. It can be stated that these variations should have influence on the results presented but it has been discovered that the hotels that offered information on their respective energy consumption has been quite representative of the generally acceptable energy demand. Hence, these flaws do not pose any serious implication on the final output of this work but these areas are still receiving attention for the sake of further investigation.

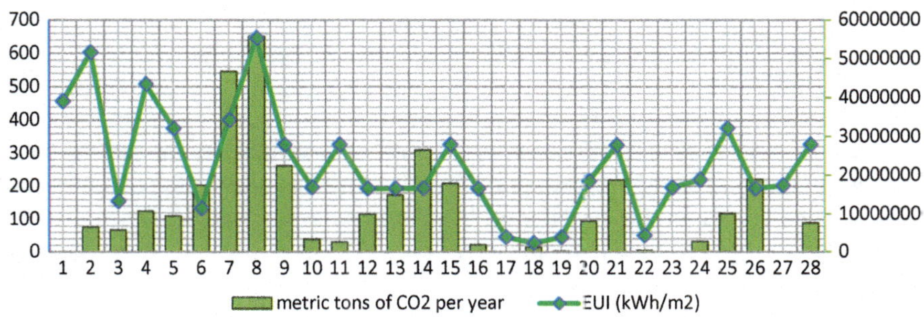

Fig. 15 Carbon emission and EUI of each of the 28 hotel buildings

Table 10 Correlations between CO_2 emission and the energy use indexes of 28 sampled hotels

		Energy consumption per unit guestroom	Energy consumption per unit worker	Energy consumption per unit equivalent guestroom	Energy consumption per unit area
	Sample size	28	28	28	28
CO_2 emission	Pearson correlation	0.561**	0.394*	0.549**	00.09
	Sig. (2-tailed)	0.002	0.038	0.002	0.65

*Correlation is significant at the 0.05 level (2-tailed)
**Correlation is significant at the 0.01 level (2-tailed)

Since Lagos is the commercial hub of Nigeria; hence the research herein developed could, as well, be applicable to other parts of the country.

Thus, the Correlation analysis was also carried out between the emitted CO_2 from the hotels buildings and normalized energy consumption. This was analysed in order to specify a value reflecting the level of CO_2 emission in relation to the normalized energy consumption. Statistical methods such as mean, median, percentile and mode were also employed so as to establish a normalized energy consumption benchmark. Furthermore, in order to find the energy use indexes responsible for the CO_2 emissions, correlation analysis was carried out and the result is as displayed in Table 9. Meanwhile the results show that the energy consumption per unit guestroom and energy consumption per unit equivalent guestroom have significant correlations with CO_2 emission. Thus, the nexus of the Statistical options for benchmarking annual energy consumption per unit guestroom and annual energy consumption per unit equivalent guestroom are as established in Table 10.

It must be stated that since a chunk of the energy is supply by the captive generators; the value stated would grievous impact on the energy balance of the environment. This gives the connection of the energy to emission of the industry.

6 Conclusion

In order to establish an electrical energy consumption benchmark for the hotel industry in Nigeria, energy consumption data for varieties of 28 different categories of star rated hotels in Lagos Nigeria has been collected and analysed using a number of well tested standard statistical tools. Thus, various standard physical/engineering factors that can influence energy consumption were considered. In which case the information regarding the Operational, building physical characteristics and building energy use for a number of 5-, 4-, 3- and 2-star hotels was collected. Based on the application of the standard statistical tools, the energy use intensity quota is established and mathematically expressed as $45.61\,\mathrm{Wh/m^2} \leq \mathrm{EUI} \leq 338.35\,\mathrm{kWh/m^2}$. Hence the resulting correlation analysis depicts that the total floor area is the most correlated in relationship with the annual energy consumption; this is thus closely followed by the number of workers. Meanwhile, an examination of the emission analysis shows a high CO_2 emission by the hotels which suggests that the industry would be contributing quite a quantum of greenhouse gas to the atmosphere.

From the foregoing, therefore, the energy use indexes responsible for the CO_2 emissions was also analysed using correlation analysis, the results show that the energy consumption per unit guestroom and energy consumption per unit equivalent guestroom were the most correlated with CO_2 emission. The suggestion and recommendation is that there should be installation of carbon sequestration and capturing tools in the retrofitting of the hotel's energy facility system to reduce the amount of carbon contents in the atmosphere due to the type of energy source being implemented. This means that the carbon dioxide meter should be installed in the facility to assist in providing warning signs when the carbon content is overshooting the maximum permissible level. For now, these facilities are not available in most of the hotels in the country, probably due to the fact that there is no deliberate penalty for carbon emission among the country's energy intensive industries. Thus, it is highly suggested that there should moderation of carbon emission by the Federal Environmental Protection Agency (FEPA).

Moreover, though energy benchmarking protocol and CO_2 emission analyses have herein been carried out in the Nigeria's Hotels using Lagos as a pilot study, there is still room for improvements on energy consumption in the hotel industry. Future research will be conducted on the sustainability of energy efficient and conservation methods in the hotel sector. For example, energy management measures based on multi-criteria decision-making (MCDM) which involves key stakeholders (facility managers, investor and government) need to be carried out. Such studies would determine the most suitable maintenance strategy and energy efficient measures for facilities by considering the four main maintenance strategies (preventive, corrective, condition-based and predictive). The selection process will be based on sustainability criteria STEEP (Social, technical, economic, environmental, policy).

Acknowledgements The authors immensely appreciate the Department of Electrical/Electronics Engineering, University of Lagos, Nigeria for critical assistance offered as well as the staff of Hotel and Catering Management of Yaba College of Technology, Lagos Nigeria for information provided as well as the past and current Masters students of the corresponding author who offered Energy Resources and Management course (taught by him) for their critiques/contributions, especially Mr. Oluwabori Babatunde. The valuable assistance of Dr. Ighravwe Desmond for making available some of the software applied in this work is also highly commended.

References

Abomeh OS (2012) Hospitality and tourism manpower training and education in Nigeria. Afro Asian J Soc Sci 3(3.4):1–18

Ali Y, Mustafa M, Al-Mashaqbah S, Mashal K, Mohsen M (2008) Potential of energy savings in the hotel sector in Jordan. Energy Convers Manag 49:3391–3397

Amadi O (2008) Nigeria hotels and boom in tourism. Businessday Newspaper, p 9

Anderson RI, Fok R, Scott J (2000) Hotel industry efficiency: an advanced linear programming examination. Am Bus Rev 18(1):40–48. Available at: http://search.proquest.com/docview/216312569?accountid=14744

Becken S, Frampton C, Simmons D (2001) Energy consumption patterns in the accommodation sector—the New Zealand case. Ecol Econ 39(3):371–386

Bohdanowicz P, Churie-Kallhauge A, Martinac I, Rezachek D (2001) Energy-efficiency and conservation in hotels—towards sustainable tourism. In: 4th international symposium on Asian Pacific Architecture, Hawaii, April, pp 1–12

Bohdanowicz P (2005) Sustainable hotels—environmental reporting according to Green Globe 21, Green Gloves Canada/GEM UK, IHEI Benchmarkhotel and Hilton Environmental Reporting

Bohdanowicz P, Martinac I (2007) Determinants and benchmarking of resource consumption in hotels—case study of Hilton International and Scandic in Europe. Energy Build 39:82–95

Chan WW, Lam JC (2002) Prediction of pollutant emission through electricity consumption by the hotel industry in Hong Kong. Int J Hospitality Manag 21(4):381–391

Chung W, Hui YV, Lam YM (2006) Benchmarking the energy efficiency of commercial buildings. Appl Energy 83(1):1–14

Cibse T (2006) Degree-days: theory and application. Chartered Institution of Building Services Engineers, London

Deng S (2003) Energy and water uses and their performance explanatory indicators in hotels in Hong Kong. Energy Build 35:775–784

Deng S, Burnett J (2000) A study of energy performance of hotel buildings in Hong Kong. Energy Build 31:7–12

Duarte R, De Matos M (2006) Energy benchmarking of double-skin façade buildings

European T.C. of the & Energy (2001) Energy savings by combined heat cooling and power plants (CHCP) in the hotel sector. Stockholm, Sweden

Farrou I, Kolokotroni M, Santamouris M (2012) A method for energy classification of hotels: a case-study of Greece. Energy Build 55:553–562

Farrou I (2013) Investigation of energy performance and climate change adaptation strategies of hotels in Greece. Ph.D. thesis, Brunel University

Filimonau V, Dickinson J, Robbins D, Huijbregts MAJ (2011) Reviewing the carbon footprint analysis of hotels: life cycle energy analysis (LCEA) as a holistic method for carbon impact appraisal of tourist accommodation. J Clean Prod 19(17–18):1917–1930

Hong T, Yang L, Xia J, Feng W (2014) Building energy benchmarking between the United States and China: methods and challenges. In: Proceedings of the 8th international symposium on heating, ventilation and airconditioning, vol 3, building simulation and information management, Lecture Notes in Electrical Engineering, 263, pp 473–486

Hui SCM, Wan MCW (2013) Study of hotel energy performance using data envelopment analysis. In: 12th International Conference on Sustainable Energy technologies (SET-2013), pp 26–31

Hwang SN, Chang TY (2003) Using data envelopment analysis to measure hotel managerial efficiency change in Taiwan. Tour Manag 24(4):357–369

Ibeh PE, Ikegwu EM, Amaechi CJ (2016) Tourism development: a revolutionalised sector as a panacea for dwindling hospitality business in Nigeria. Int J Sci Soc Yabatech 4(1):39–46

International Tourism Partnership (2014) Environmental management for hotels: the industry guide to sustainable operation. International Tourism Partnership, pp 1–23

Keh HT, Chu S, Xu J (2005) Efficiency, effectiveness and productivity of marketing in services. Eur J Oper Res 170(1):265–276

Khemiri A, Hassairi M (2005) Development of energy efficiency improvement in the Tunisian hotel sector: a case study. Renew Energy 30(6):903–911

Krese G, Prek M, Butala V (2012) Analysis of building electric energy consumption data using an improved cooling degree day method. Strojniški vestn—J Mech Eng 58(2):107–114. Available at: http://en.sv-jme.eu/data/upload/2012/02/05_2011_160_Krese_05.pdf

Li Q, Ren P, Meng Q (2010) Prediction model of annual energy consumption of residential buildings. In: 2010 international conference on advances in energy engineering (ICAEE), pp 223–226

Li Z, Han Y, Xu P (2014a) Methods for benchmarking building energy consumption against its past or intended performance: an overview. Appl Energy 124:325–334. Available at: http://dx.doi.org/10.1016/j.apenergy.2014.03.020

Li Q, Xiong W, Chen J, Zheng G, Li Z, Mills JL, Liu A (2014b) A robust test for quantitative trait analysis with model uncertainty in genetic association studies. Stat Interface 7:61–68

Lu S, Wei S, Zhang K, Kong X, Wu W (2013) Investigation and analysis on the energy consumption of starred hotel buildings in Hainan Province, the tropical region of China. Energy Convers Manag 75:570–580

Lung-Tan L (2015) Performance and efficiency evaluations of international hotels in Taipei. Int J Econ Finance 7(7):64–73

Ma JH (2005) Foundation of management and statistic. Tianjin University Press, Tianjin

MacDonald JM (2001) Benchmarking residential energy use, http://amet-me.mnsu.edu/userfilesshared/solarwall/Benchmarking/Benchmarking%20Residential%20Energy%20Use.pdf/. Accessed 12 Jan 2018

Manasakis C, Apostolakis A, Datseris G (2013) Using data envelopment analysis to measure hotel efficiency in Crete. Int J Contemp Hospitality Manag 25(4):510–535

Maxwell KD, Forselius P (2000) Benchmarking software development productivity. IEEE Softw 17(1):80–88

Min H, Min H, Joo SJ, Kim J (2008) A data envelopment analysis for establishing the financial benchmark of Korean hotels. Int J Serv Operat Manag 4(2), 201. Available at: http://www.inderscience.com/link.php?id=16611

Morey RC, Dittman DA (1995) Evalatn a Hotel GM's performance: a case study in benchmarking Cornell Hotel Restaurant Adm Q 36(5):30–35

Oluseyi PO, Babatunde OM, Babatunde OA (2016) Assessment of energy consumption and carbon footprint from the hotel sector within Lagos, Nigeria. Energy Build 118:106–113

Önüt S, Soner S (2006) Energy efficiency assessment for the Antalya Region hotels in Turkey. Energy Build 38(8):964–971

Oukil A, Al-Zidi A (2017) Benchmarking the hotel industry in Oman through a three-stage DEA. SQU J Humanit Soc Sci

Papamarcou M, Kalogirou S (2001) Financial appraisal of a combined heat and power system for a hotel in Cyprus. Energy Convers Manag 42(6):689–708

Pereira Oliveira RSL, Pedro MIC, da Cunha Marques RDR (2015) Efficiency evaluation of Portuguese Hotels in the Algarve using data envelopment analysis (DEA). Rev Bras de Gestão de Negócios 17(54):788–805

Pieri SP, Santamouris M et al (2015) Identifying energy consumption patterns in the Attica hotel sector using cluster analysis techniques with the aim of reducing hotels CO_2 footprint. Energy Build 94:252–262

Poldrugovac K, Tekavcic M, Jankovic S (2016) Efficiency in the hotel industry: an empirical examination of the most influential factors. Econ Res-Ekonomska Istraživanja 29(1):583–597

Priyadarsini R, Xuchao W, Eang LS (2009) A study on energy performance of hotel buildings in Singapore. Energy Build 41(12):1319–1324

Reddy TA, Saman NF, Claridge DE, Haberl JS, Turner WD, Chalifoux AT (1997) Baselining methodology for facility-level monthly energy use-part 1: theoretical aspects. ASHRAE Trans, pt 2 ed., 103:336–347

Rossello-Batle B, Moia A, Cladera A, Martinez V (2010) Energy use, CO_2 emissions and waste throughout the life cycle of a sample of hotels in the Balearic Islands. Energy Build 42(4):547–558

Ryu J-H, Hong W-H, Seo Y-K (2014) A study of the analysis on the properties of electricity peak load of large hotel building in consideration of energy efficiency. Adv Sci Technol Lett 47 (2):6–10

Sanni MR (2009) The influence of the economy on hospitality in Nigeria. Ethiop J Environ Stud Manag 2(1):29–34

Shiming D, Burnett J (2002) Energy use and management in hotels in Hong Kong. Hospitality Manag 21:371–380

Sun H, Lee SE, Priyadarsini RMT, Wu X, Chia Y, Majid H-S (2006) Building energy performance benchmarking and simulation under tropical climatic conditions. International workshop of energy performance environmental quality of buildings, Milos Island, Greece, pp 1–9

Takane Y, Yanai H, Hwang H (2006) An improved method for generalized constrained canonical correlation analysis. Comput Stat Data Anal 50(1):221–241

Tantiwanit K (2007) Establishing energy consumption benchmarks of office buildings in Bangkok

Teng C, Horng J, Hu M, Chien L, Shen Y (2012) Developing energy conservation and carbon reduction indicators for the hotel industry in Taiwan. Int J Hospitality Manag 31(1):199–208

Trung DN, Kumar S (2005) Resource use and waste management in Vietnam hotel industry. J Clean Prod 13:109–116

Tsai H (2009) Star-rated hotel productivity in china: a provincial analysis using the DEA cross-efficiency evaluation approach. J China Tourism Res 5(3):243–258

Tzschentke NA, Kirk D, Lynch PA (2008) Going green: decisional factors in small hospitality operations. Int J Hospitality Manag 27(1):126–133

Wang JC (2012) A study on the energy performance of hotel buildings in Taiwan. Energy Build 49:268–275

Wang JC, Huang K-T (2013) Energy consumption characteristics of hotel's marketing preference for guests from regions perspective. Energy 52:173–184

Wang L (2005) Support vector machines : theory and applications

Wikipedia, Correlation coefficient. https://en.wikipedia.org/wiki/Correlation_coefficient. Accessed 6 Nov 2017

Wu J, Song H (2011) Operational performance and benchmarking: a case study of international tourist hotels in Taipei. Afr J Bus Manag 5(22):9455–9465

Xin Y, Lu S, Zhu N, Wu W (2012) Energy consumption quota of four and five star luxury hotel buildings in Hainan province, China. Energy Build 45:250–256

Xuchao W, Priyadarsini R, Siew Eang L (2010) Benchmarking energy use and greenhouse gas emissions in Singapore's hotel industry. Energy Policy 38(8):4520–4527

Xuchao W (2007) Energy performance of hotel buildings. Thesis, National University of Singapore

Xydis G, Koroneos C, Polyzakis A (2009) Energy and exergy analysis of the Greek hotel sector : an application. Energy Build 41:402–406

Yao Z, Zhuang Z, Gu W (2015) Study on energy use characteristics of hotel buildings in Shanghai. Procedia Eng 121:1977–1982

Yen FL, Othman M (2011) Data envelopment analysis to measure efficiency of hotels in Malaysia. SEGi Rev 4(1):25–36

Yik FWH, Burnett J, Prescott I (2001) Predicting air-conditioning energy consumption of a group of buildings using different heat rejection methods. Energy Build 33(2):151–166

Yoon S-H, Park C-S (2017) Objective building energy performance benchmarking using data envelopment analysis and Monte Carlo sampling. Sustain (Switzerland) 9(5)

Zhao H, Magoulès F (2012) A review on the prediction of building energy consumption. Renew Sustain Rev 16:3586–3592